SCOTOPHOBIN

Darkness at the Dawn of the Search for Memory Molecules

Louis Neal Irwin

Hamilton Books
A member of
The Rowman & Littlefield Publishing Group
Lanham · Boulder · New York · Toronto · Plymouth, UK

Copyright © 2007 by
Hamilton Books
4501 Forbes Boulevard
Suite 200
Lanham, Maryland 20706
Hamilton Books Acquisitions Department (301) 459-3366

Estover Road
Plymouth PL6 7PY
United Kingdom

Library of Congress Control Number: 2006930851
ISBN-13: 978-0-7618-3580-6 (paperback : alk. paper)
ISBN-10: 0-7618-3580-6 (paperback : alk. paper)

In memory of

Georges Ungar
and
Frederick E. Samson, Jr.

Table of Contents

List of Tables

Foreword

Neuroscience is the interdisciplinary field that includes all the sciences, from sub-molecular biology through psychology and psychiatry, whose practitioners are involved, one way or another, in trying to understand how the brain and mind "works". It is a relatively new field, barely 50 years old, and it has grown from a relatively small group of a few hundred scientists in the 1960s to a large, wide-ranging field of many thousands, including over 35,000 members of the Society for Neuroscience in North America alone, with many other neuroscience organizations throughout the world.

Many of the neuroscience pioneers are still with us, including the author of this fascinating history of some of the events in the development of the field. In a series of chapters filled with the talented and sometimes quirky personalities involved in some of the landmark events in the early history of this new science, Lou Irwin chronicles his own experiences as a participant and observer. In particular, he tells the inside story of the early exciting and often frustrating and controversial efforts to solve the mystery of memory and learning—how the brain processes new information and how that information is retained as memory. In this "learning and memory" aspect of early neuroscience, much of the research focused on the transfer of learning: Can neurobiological changes in the brain, pre- and post-learning, be located and identified; and, if so, can this changed-brain information be "transferred" from the learned brain to the unlearned brain of a naïve recipient? Indeed, transference seemed to work in many cases. Recipient animals appeared to learn more easily the behavior of the donor animal. How and why this worked was anyone's guess. Were memory molecules involved?

This probably overly simplistic approach to studying the enormously complicated phenomena of learning and memory, nevertheless gave neuroscience a strong impetus and also much publicity, attracting new researchers into neuroscience and inspiring new strategies and novel experimental designs as well as new animal models. One of these new models, which, perhaps, by its very strangeness, attracted much interest and brought many new people into the field (including Nobel Laureate Melvin Calvin and some of his associates) is described by Irwin at some length. It involved the planarian, a simple worm that could be trained to turn in a particular direction. When trained worms were fed to untrained worms, the naive worms were seen to move more readily in the way the trained worms did. Inspired by the publicity his planarian findings attracted, James McConnell of the University of Michigan, the prime planarian "worm runner", was able to attract increased research support. He even started to publish a new journal, the *Worm Runner's Digest*, which published the results of the planarian research – but also included a humor section. After all, how could one not see the humor in this almost ludicrous experimental paradigm? As a young biochemistry graduate student at the University of Kansas, Lou Irwin studied the different kinds of research going on in the field and was able to participate with his own memory studies as a summer research assistant in the laboratory of the late Georges Ungar at the Baylor College of Medicine in Houston. He had carefully followed the results of the experiments in many labs involving the biochemical brain changes accompanying learning. Especially Inspired by Ungar's findings of the transfer of sound habituation and morphine tolerance in rats, as well as the research of Holger Hydén in Sweden who reported finding RNA changes in the brains of rats trained to climb a wire, he proceeded to do his own experiments involving conditioning rats and the transfer of learning. Initially he was, remarkably, able to demonstrate positive results. But, as with many (but not all) other researchers at the time, he could not repeat his positive results with a new set of rats, and had to return to Kansas before the inconsistencies could be resolved.

Lou's mentor at the University of Kansas was the late Fred Samson, and Lou gives his honored teacher a fitting tribute, sentimental and venerating, but with a realistic portrait of the personal quirks and foibles of this talented man. Fred Samson, an unusually warm and caring person, had a phenomenal career, ranging from osteopathy to show business performer as an acrobat, to army medic on Guadalcanal. Following the war and graduate training in physiology, Samson went on to a successful research and teaching career at the University of Kansas. But his world was changed by a sabbatical year at MIT's Neurosciences Research Program (NRP). This was an organization founded by the late Francis O. Schmitt, Head of Biology and Institute Professor at MIT, which was devoted to the investigation of how the mind/brain works—how the brain controls behavior, including mental activity. This experience changed Samson's research focus and brought him fully into the new field of neuroscience. As one of NRP's first Staff Scientists, Samson worked closely with Schmitt and other staff members in planning and organizing the bimonthly Work Sessions, subject-oriented meetings of specialists in particular areas of neuroscience, which often opened up new areas of research. One of these was a Work Session on "Axoplasmic Transport", in which Samson worked with Schmitt, Paul Weiss, Sam Barondes, and NRP Staff members to organize and then help write the NRP Bulletin largely responsible for opening up a whole new field of research on this intrinsic and poorly understood activity of the neuronal axon. Lou later followed his mentor to the NRP and, as a Staff Scientist, became a valuable asset to this organization. His reminiscences about "settling in" to the unusual environment of this often high-pressure "think tank" are sometimes humorous, but are also very poignant. He paints realistically his experiences as a southerner adjusting to the often severe New England weather and to the duties of an organization, where one day might be spent discussing science with a Nobel Laureate, while the next was devoted to xeroxing and editing a poorly written manuscript.

This is indeed a fascinating book, reviewing some of the important events in the earlier history of neuroscience, while recounting the trials and tribulations of some of the important players, including their controversies and disputes (often colored with personal animosities) over scientific issues, all in the on-going effort to make sense of the enormous complexities of the brain.

One will come away from this book with a feeling for the excitement of this field and of the pleasure, even joy, that some—perhaps all—neuroscientists experience in carrying out their work. Lou Irwin has succeeded in communicating vividly in this book a sense of the excitement and importance of this area and the hope that, perhaps in our time, we will be able to answer the question that philosophers and thinkers, since the earliest days of our intellectual history, have struggled with—where is mind, where is consciousness, where is memory and how does the brain work to control all this multidimensional inner and outer world?

George Adelman, Co-Editor
Encyclopedia of Neuroscience

Preface

This is the true story of an episode in the development of neuroscience during the 1960s and 1970s, told at three levels.

It is first a factual history of the discovery of neuroactive peptides, with primary focus on the strategy of whole-animal assays for detecting behaviorally-induced changes—the so-called "behavioral transfer" paradigm, now generally regarded as an early, overreaching and unsuccessful attempt to understand the molecular basis of memory.

At a second level, it is a personal account of how the story unfolded from my perspective, as a young scientist caught between the contending influences of two powerful mentors, one of whom despised the other, both of whom I sought to honor.

At the third level, it seeks to provide an uncritical observation of the way conventional science deals with unconventional challenges. At base, it shows how the scientific process generally moves toward a closer approximation to the truth over the long run.

The narrative of the story is written for the lay reader, with a little more technical explanation than the expert will need, and without documentation. However, an essay recounting the history of behavioral transfer research and the backdrop against which it unfolded, is provided at the end, complete with bibliographic references. For readers who prefer to focus on the factual story itself, the addendum stands alone.

Because this is a true story, I have tried to be as accurate as possible. My own detailed notes and correspondence have been supplemented by personal interviews with Mark Abel, George Adelman, Bernard Agranoff, Samuel Barondes, Robin Barraco, Edward Bennett, Floyd Bloom, Rodney Bryant, Bill Byrne, Dennis and Nancy Dahl, Dominic Desiderio, Adrian Dunn, Arnold Golub, Avram Goldstein, Dianna Johnson, David Malin, James McConnell,

James McGaugh, Mark Rosenzweig, Fred Samson, Richard Thompson, Alberte Ungar, and John Wilson.

The manuscript was read by George Adelman, Bernard Agranoff, Samuel Barondes, Edward Bennett, Adrian Dunn, Mark Rosenzweig, Larry Stern, and, in an earlier version, by Mark Abel, Donna Byers, Dan Michael, Dianna Johnson, and Fred Samson. Sean Stewart, Karen Talentino, Sandra Williams, and Judith Wittenberg critiqued a few early chapters. All made helpful comments and suggestions, which I truly appreciate. Larry Stern was particularly helpful and generous in sharing the benefit of his extensive knowledge about the history of behavioral transfer research. Much of the material on the discovery of endogenous opiates was based on the excellent account by Jeff Goldberg in *Anatomy of a Scientific Discovery.* Some errors surely have remained. I accept responsibility for those, and invite corrections.

My wife, Carol, lived much of this story with me, and supported me without reservation through the ups and downs of a bumpy career. I can't thank her enough for enduring the ride and making a happy home for our family, while pursuing two careers of her own.

El Paso, Texas
January, 2006

Epiphany

Christmas, 1965

On the morning of December 22, 1965, I stood beside a refrigerator in a laboratory at the Baylor College of Medicine in Houston, holding in my hand a test tube containing a powdery extract from the brain of a rat. My host, Dr. Georges Ungar, had removed it from the freezer and handed it to me with care, as it represented what he hoped would lead to the culmination of his distinguished career. The extract was a mixture of substances that included, he felt, the molecules that coded the memory of the behavioral choices the rat had been trained to make to run a maze.

I stared at the indistinctive powder, aware from my training in biochemistry of the tremendous capacity for information that molecules possess, far below the level the eye can perceive. But could a molecule hold the memory of past behavior and experience? Were memory molecules really like snippets of magnetic tape bobbing about in the brain, erasable from one behavior to the next, readable through the reminiscent machinery that the neural circuitry of the brain provides?

Why not? Was the notion so preposterous? The brightest minds in psychology had been searching for the secret of memory for over half a century with no discernible success. Karl Lashley, after 30 years of monumental effort, concluded in whimsical exasperation that memory appeared to be an impossibility. But memories are very much a reality; and if our minds have a material basis, it follows that memories derive from information that the brain can store for a lifetime in a tangible form.

Georges Ungar, a Hungarian-born Frenchman with a career of research on the chemical basis of cellular responses to stimulation mostly behind him, had no difficulty believing that the brain responds to stimulation by producing molecules that endure. Only twelve years earlier, James Watson and Francis Crick had proposed a structure for deoxyribonucleic acid (DNA), the most enduring of the cell's molecules and the repository of genetic information that encodes the multiplicity of information that a cell has to inherit to carry out all its genetically programmed functions.

As a first-year graduate student with a background in both chemistry and psychology, the notion of memory molecules floating about the brain struck me as a bit oversimplified. But to the extent that every organism is ultimately a collection of molecules, I had to assume that the information that an animal stores when it learns something has a chemical reality at some level. The possibility that I might be looking at the very molecules involved in the storage of behavioral information slowly sank in. Then, like a spark that kindles into a flame, it fired my imagination. Slowly at first, then with a rush, a sense came over me that I was standing at a time and place of historical importance—overriding historical importance—in my life. It was one of those profoundly satisfying moments that we all have on rare occasion, in which we can't imagine a place or person we would rather be.

The story that unfolded after that morning did not have a triumphant conclusion. Several molecules were isolated—scotophobin being the best known—that were alleged to have behavioral effects. Other peptides were shown to be neuroactive beyond the shadow of a doubt. And I did go on to have a modestly successful and rewarding career in science. But scotophobin's discoverer, Georges Ungar, died in greater ignominy and less honor than his considerable accomplishments warranted. Scotophobin was discredited but never fully explained. And those of us who worked even briefly on the disreputable notion of biochemical transfer of memory, found it a ride we would never forget.

This book is my personal account of that ride. I believe it shows some important points about the practice and sociology of science, but I'm not sure what those points are. I'm still coming to grips with the consequences of that morning beside the refrigerator in Georges Ungar's lab. But I trust that history's verdict on the worth and significance of scotophobin in the long run will be accurate. That is the beauty of the scientific endeavor. This account is intended to document and explain the early chapters of that history from one person's perspective. I believe that the final chapter is yet to be written.

Prelude

Which Twin Has the Memory?

James Vernon McConnell's career as a disc jockey was going nowhere in the summer of 1947. The day after his graduation from Louisiana State University, he had been hired by WTPS in New Orleans on the strength of his experience as a part time announcer at the L.S.U. station in Baton Rouge. Despite a strong voice and mastery of the mechanics of broadcasting, his southern accent was too thick and deep for the management of a major metropolitan radio station to tolerate. Just weeks after he was hired, he expected to be fired, and he didn't know what he would do next.

He had intended originally to be a chemical engineer. Born in Okmulgee, Oklahoma in 1925 but raised in Shreveport, he had decided to capitalize on the opportunities of the sugar cane industry in Louisiana. He enrolled at L.S.U. in 1942 at the age of 16, too young for military service. It took only a semester for him to discover—ironically, as it would later turn out that chemistry was not for him. He found that he enjoyed drama, and with few males around, he was eagerly sought for acting and announcing roles on campus. He volunteered for the Navy in 1943 but wasn't inducted till 1944, and by the time he completed officer's training, the war was over. He was a crew-member of the ship that ferried the atomic bomb to its Bikini Island test site in the Pacific, but saw no real excitement other than the first test blast. He returned to L.S.U. in 1946 and enrolled as a psychology major. While biding his time at home in Shreveport prior to his induction, he had taken an introductory psychology course at Centenary College. Given his strict and prudish upbringing, the open and frank discussions of Freud, sex, and the scientific study of human behavior were an exciting revelation. Chemistry was a pale and boring subject, by comparison.

Now with a college degree in psychology but no real focus other than radio, he expected to be fired from his first job. The management at WTPS referred to it, however, as a "reassignment." He was farmed out to a smaller station in Lake Charles, where the degree of his Southern drawl didn't matter. There he met Rosa Hart, from the cultural side of town, who persuaded him that he would never go anywhere with a thick Southern accent. Under her tutelage, McConnell acquired a non-descript "mid-Atlantic" pronunciation style. In the course of retraining his voice, Rosa Hart opened his eyes to a larger world of literature, history, and culture that built his confidence and gave him the beginning of wisdom. He moved to Galveston, where he became a station manager, then was hired by WLW-TV in Cincinnati as a writer.

Television was new and experimental in 1950. WLW radio had 12 writers. McConnell was the only writer for WLW-TV, so he wrote everything: commercials, skits, an original musical comedy every week, and the first day-time soap opera, which the management decided not to run on grounds that soaps would never work on television. Expanding on the real education that he had begun in Lake Charles, McConnell learned more about life and a lot about writing; but the pace was grueling and he got an ulcer. Then the death of his father in March, 1951, gave him a reason to return to Shreveport and a less hectic way of life. Someone was needed to run the family business, a bus station cafe. But the over-stimulation of Cincinnati was exchanged for a tedious job devoid of the intellectual challenge to which he had become accustomed, so he sank into a deep rut and got another ulcer. A friend told him one day, "You're rotting. You've got to do something. . . . Go get a Ph.D."

By the time McConnell decided to do something, like get a Ph.D., it was August of 1951, and the only graduate school governed by state laws and admissions criteria lenient enough to admit him for the fall semester was the University of Texas in Austin. At that time, the practice of the Psychology Department, which was one of the best in the South, was to admit large numbers of students and flunk out a sizeable proportion of them within a year. McConnell was not among the failures. He enjoyed graduate study and became one of the department's star students. He enrolled with the vague notion of becoming a clinical psychologist, and made straight As in his courses. At the end of his first year, though, Wayne Holtzman—developer of the quantitative ink blot test for personality evaluation and director of the clinical program—told McConnell that he had to leave the program because his "attitude" was bad.

"I asked him, 'How can you do that? I have good grades.' " McConnell recalled.

"Yes, but every time one of your teachers writes something on the board, you stick up your hand and say 'What are the data?' "

That, according to McConnell, was Holtzman's definition of a bad attitude.

McConnell by that time was already leaning toward experimental psych-ology, so Holtzman's counseling simply hastened his transition to an area of psychology where data did not present an attitude problem. At first, McConnell worked with Jeff Bitterman, a comparative psychologist who had come from Emory University where he had worked with chimpanzees. But finding no chimps at the University of Texas ("save for those on the staff," according to McConnell), Bitterman had taken up the study of behavior in earthworms. It was at this point that McConnell became friends with another of Bitterman's grad-uate students, Robert Thompson.

McConnell and Thompson started thinking about worms. Thompson had read Donald Hebb's 1949 book on *The Organization of Behavior*, which postulated that learning occurs because of functional changes at the synapses of the nervous system. The synapse is the point of connection between different nerve cells (neurons). When a neuron is activated, it conducts a tiny wave of electrical current from the point of activation to the end of a long extension, the axon, which protrudes from the swollen bulk of the cell body (where the nucleus and a lot of other metabolic machinery is located). At the end of the axon, where it comes in contact (or, to be precise, in near-contact) with a branch or the cell body of another cell, the electrical wave may either jump the gap like a spark between two wires or cause the release of molecules (called neurotransmitters) that diffuse across the tiny synaptic gap and reactivate the electrical wave in the next ("postsynaptic") neuron.

The nervous system, at its simplest level, is a collection of neurons arranged through synaptic interconnections into a complex three-dimensional matrix. As the traveling waves of electrical excitation move from one neuron to the next, transmitted at their point of interconnection either electrically or by chemical neurotransmitters, they generate a particular sequence or pattern of activation. Hebb proposed, first of all, that a unique pattern of activation (a particular "cell assembly") is the way that the nervous system represents unique information, such as the sound of a bell or the taste of meat. Secondly, Hebb reasoned that the relationship between different types of information is represented by activating specific cell assemblies in a particular order, or "phase sequence." Thirdly, he assumed that if one cell assembly (representing, say, the sound of a bell) is consistently activated with another cell assembly (say for the taste of meat), the two bits of information will become permanently associated, because the first cell assembly will acquire the ability to consistently activate the second cell assembly.

Since the neurons of one cell assembly interconnect with those of other cell assemblies through synaptic connections, it stands to reason that a functional change must occur in the synaptic connection between the two cell assemblies to make the first capable of repeatedly and consistently triggering the second. Thus the ability of an animal to learn to associate the sound of a bell with the approach of food boils down to changes in the efficacy of synaptic connections in its nervous system. What the changes might be was left vague by Hebb. He considered that an enlargement or growth process at the synapse was the most likely mechanism, but conceded that long-lasting metabolic changes or chemical modifications below the level of structural alterations were also possible.

The basic anatomy of nerve cells and synapses has been known for more than a century, and for almost that long students of the nervous system have assumed that changes in learning involve changes at the synaptic connections. When Hebb wrote his book, and when Thompson read it, there was no justification for this assumption other than the rather compelling logic that the point of contact between neurons is the most plausible switch point for modifying their functional relationship. Thompson and McConnell were clearly aware that the "Hebbian" (modifiable) synapse was a theoretical concept without experimental support. Hebb himself referred to it as "a neurophysio-logical postulate."

If learning requires a functional change at a synapse, then learning should be possible only in organisms that have a nervous system with synaptic connections. Some organisms without such nervous systems are clearly capable of behavior. The ameba, a one-celled protozoan familiar to any biology student, can move about, respond to stimuli such as touch or light, and show other elements of behavior; but it has no nervous system at all. If Hebb was right it should not be able to learn. A more complex organism, but one with the superficial appearance of behavior not much more complicated than that of the ameba, is the flatworm, or planarian, typically found on the underside of rocks in freshwater ponds and streams. Planaria have flattened, spade-shaped bodies a few millimeters long, with a couple of cross-eyed light-sensitive spots at one end, just in front of a concentrated mass of nerve cells that zoologists charitably call a brain. The nervous system consists of true neurons, joined by synaptic connections essentially similar to those of more complex animals. Planaria are the simplest, most primitive animals to have such an organized, synaptic nervous system. If Hebb's "postulate" was correct, planaria had to be capable of learning. This is what Thompson and McConnell set out to show, in McConnell's kitchen, in 1952.

McConnell and Thompson began running flatworms down a foot-long plastic trough filled with water in contact with electrodes at either end. When the animal was gliding smoothly in a straight line, a light above the trough would be turned on, then three seconds later an electrical current would be activated. The shock of the current caused an instinctive contraction of the worm's body—an unlearned (unconditioned) behavior. At first, the light had no effect, but with repeated trials, some worms began to contract at the onset of the light before they received the shock. This, according to the experimenters, was learned (conditioned) behavior, since it depended on the experience of a temporal association between the unconditioned stimulus (shock) and a conditioned (light) stimulus. It was a straightforward extension of the classical conditioning technique used by Pavlov in teaching dogs to salivate in response to otherwise neutral stimuli associated with food.

The experiments were conducted in McConnell's kitchen, because Bitterman didn't want them done in his lab. At that time, Jeff Bitterman was an "operant purist," meaning that he distrusted the study of any form of behavior that could not be measured by automated instrumentation. The sometimes subtle head-turns and contractions that constituted "correct" responses in the worm runner's trough did not lend themselves easily to automation. A long-standing dilemma in experimental psychology is whether the objectivity gained by automating behavioral measurements is more important than the richness and subtlety that can be detected by the human but subjective eye.

For the two graduate students in Jeff Bitterman's lab, the dilemma was political. It was particularly political in that department at that time, as

Psychology was badly split into factions (not an unusual situation for academic departments, especially those of psychology). The only way to get through such a minefield, McConnell concluded, was to attach himself to a professor with clout who could clear the way. Among other things, this meant carrying out a doctoral research program of interest to the sponsoring professor. With Bitterman uninterested in flatworms, and the other faculty members focusing on the traditional concerns of psychology departments—white rats and college sophomores—McConnell turned away from classical conditioning of planaria. He attached himself instead to Karl Dahlenbach, chairman of the department, and carried out a study on "Aftereffects of Rotation in the Visual Environment," using human subjects.

He wrote his dissertation in Norway as a Fullbright Scholar during the 1954-55 academic year (where he "froze and went into a deep depression"), and was graduated with a Ph.D. from the University of Texas in 1956. He and Thompson never did more than the one classical conditioning experiment on planaria while in Austin. They thought the results of that one experiment were significant enough for publication, however, and the *Journal of Comparative and Physiological Psychology* agreed. The paper appeared in 1955, and was little noticed at the time.

With a doctorate in experimental psychology, Jim McConnell was hired as an Instructor by the University of Michigan in the fall of 1956. His new departmental chairman welcomed him to Ann Arbor with the reminder that Michigan was a first-class university, so to stay there McConnell would have to publish. McConnell remembers the chairman putting it this way:

> I have a favor to ask. If at all possible, try to publish good research. But if you can't publish good research, publish a lot of bad research, because the Dean won't know the difference.

Whether McConnell ended up doing good or bad research would become a subject of debate in the ensuing years. There are those who argue that in fact what McConnell—or, more particularly, some of his students—did was a lot of pretty good research, but research that gave disturbing results, and research conducted in a style that the scientific community regarded as somewhat "unseemly."

In 1956, though, the controversy lay in the future. McConnell had been told that he had to publish or perish, so he needed a topic that was readily publishable and one that he could call his own. The one subject that really intrigued him was the learning capacity of flatworms. Thompson had moved in a different research direction, apparently with no intention of returning to the running of worms (which he never did), so McConnell had the field to himself.

Having demonstrated, in his opinion, that planaria could be classically conditioned, McConnell mulled over the possibilities for his next step. His thoughts returned to a coffee break that he and Thompson had shared one day years earlier. Thompson had been thumbing through an issue of *Life* magazine when he came upon a two-page ad for Toni Home Permanents. Identical twins were pictured, one on each page with equally beautiful hair. Which twin has the home permanent, and which one had hers done at the beauty shop? asked the ad. You know, Thompson pointed out, when you cut a planarian cross-wise into a head section and a tail section, each half regenerates to give identical twins. McConnell and Thompson, with worms on their minds, began to ponder the intriguing implications. If the bisected planarian had been conditioned, which twin would retain the learned response—the one regenerated from the head, or the one that grew from the tail? It was an experiment they never got to carry out in Austin. But it was one that McConnell was now free to do.

With Allan Jacobson, an undergraduate senior honors student, and Dan Kimble, a first year graduate student, McConnell carried out the long-delayed experiment, and found that the tail-regenerates retained the conditioned behavior as well or better than the head-regenerates. Was he surprised by the results? Not really, he claimed. "It just seemed like . . . some part, some remnant of something would be in the tail." McConnell admits that his notion of why the experiment was even conducted was very imprecise at the time. "Sophistication was not my middle name. . . . I just thought it would be fun if the tail showed any retention at all. We would worry about 'how' later."

Was the experiment that launched a new way of thinking about learning in animals really done for "fun?" It wouldn't be that unusual. Philosophers of science and elementary textbooks routinely suppose that science is guided by hypothesis-testing and rigorous logic. The fact is that scientists often carry out experiments primarily because they seem like 'an interesting thing to do' at a particular point in time. Furthermore, if the experiments aren't 'fun,' at least in part at some level, they are often abandoned. The line of research ultimately undertaken may lead to a rigorous hypothesis in time. The end result may be as significant and logically defensible, but the reasons the experiments were done in the first place often have a more subjective (human, if you will) explanation.

It would be wrong to infer from the above that there was no theoretical thinking behind the transection experiment. Regeneration poses a number of classic problems in developmental biology, and transected planaria had been used for many years to try to shed light on those problems. It was known that when a planarian is cut in half, the row of cells at the border of the cut on the tail piece generate all the cells that form a new head. So cells that are part of the animal's midsection end up generating a whole new structure, including a new brain. To the extent that McConnell thought it through at the time, he was focusing on the structural aspects of the phenomenon. He and Thompson had begun the worm running experiments in search of the Hebbian synapse. The issue had therefore evolved into the question of whether a regenerating tail could 'instruct' the rebuilding of the Hebbian synapses that presumably had been altered when the (now discarded) head portion of the worm presumably had learned to respond to the light.

The source of the 'instructions' for regenerating a knowledgeable head would ultimately become the key issue. Even when learning is not involved, the phenomenon of regeneration poses an intriguing dilemma. While the cells at the cut edge of a tail piece regenerate into a body of cells that form a head, the cells just one row forward, at the back edge of the head piece, regenerate into a tail. If the cut is made one row of cells further back, cells which would have regenerated a head now regenerate a tail instead. Clearly, the information that tells a row of cells whether to regenerate a head or a tail arises from some distributed property of the transected tissue. In the case of McConnell's experiment, it appeared that information relating to behavioral experience had been stored in a distributed fashion in the aft as well as the fore compartments of the animal's body. But this *is* a more sophisticated analysis of the issue than McConnell was making at the time. Back then, he was busy simply pursuing an interesting phe-nomenon.

Could, for instance, the trained head segment of a bisected worm regenerate a tail which, when cut away from its parental head tissue regenerate a new head, and thereby a totally reconstructed animal, that retained the conditioned response? This experiment was carried out by Reeva Jacobson (no kin to Allan), an undergraduate honors student working in McConnell's lab, with positive results. To that point the experiments had been reported only through the usual form of presentation at scientific meetings and publication in professional journals which attracted no public notice. But when Reeva Jacobson's work was presented at a meeting of the American Psychological Association in September of 1959, *Newsweek* picked up the story and published a brief summary of the experiment under the title, "Strangest of Tails," complete with a diagram captioned "The Worm Sees the Light." It was the first of many reports of McConnell and his curious experiments that would appear in the popular press over the years.

Most scientists are wary of the popular press. Not that they object in principle to seeing their names in the morning paper; they just don't like to see the inevitable simplification of their work that journalism for a general readership usually requires. McConnell, however, was not so wary:

> I always was a loudmouth. I had been in radio and t.v. I knew how to get publicity. And I was certainly not reluctant to talk to the press. I didn't go hunting for publicity, but once someone would come and ask about it, I would talk to them.

For some reason—perhaps because they appeared to be simple and inexpensive yet meaningful experiments—the work on planaria captured the collective imagination of amateur scientists and high school students all across the country. McConnell was besieged with letters requesting instructions on how to carry out his experiments. It quickly became obvious that individual replies were not feasible, so he began to mimeograph a manual on the arcane science of worm running. There was something about the subject that begged for humor. And there was a lot about McConnell's personality and attitude that made it impossible for him to stick to the somber tenor of the professional journals. He had done that already in the *Journal of Comparative and Physiological Psychology*, and had nothing to prove. So he added some art work and cartoons to his manual on worm-running, inscribed it with Latin mottoes, enlivened it with liberal doses of satirical prose and poetry, and dubbed it the *Worm Runner's Digest*. The tone of irreverence that would characterize the *Digest* for the two decades of its existence was signaled in its first issue of November, 1959:

> We . . . apologize if occasional hints of levity creep into the pages of this austere scientific publication. We enjoy life, and we think working with worms can be fun as well as a worthwhile research endeavor. Our efforts in the lab are no less rigorous for our being able to joke about our results.

The *Digest* came to be perceived over time by many scientists as long on levity and short on austerity. And a surprising number of scientists, to McConnell's chagrin, could not tell the difference between the former and the latter; so he resorted to publishing the levity upside down and on the back of the pages bearing the austerity. This and related examples of frivolity came to be an issue—and to some, the main issue—even as the research itself was taking a more serious and significant turn.

By 1959, the following was known about the molecular basis of genetic information: The essence of what a cell is or does depends on the proteins that it manufactures. Proteins are large molecules that serve as either building blocks for the cell or as enzymes that facilitate chemical reactions in the cell. Proteins in the flat cells that line the surface of a leaf and produce chlorophyll are different from the proteins that give our muscle cells their long shape and ability to contract when a nerve impulse tells them to. And the reason that proteins differ from one another in function is that their three-dimensional shapes are different, just as hexagonal tiles and rectangular bricks build a different type of wall, or locks with different internal structures accept keys of different patterns.

Yet all proteins are made from the same set of about 20 different amino acids. Each amino acid has a distinct size and shape. When strung together like beads on a string, the sequence of amino acids gives the overall protein its characteristic and unique three-dimensional shape. The sequence of amino acids, in turn, is governed by the sequence of chemical bases in a long molecule called ribonucleic acid (RNA). The cell has special structures outside its nucleus for taking the ribbon-like molecules of RNA, lining up amino acids in the sequence coded by the RNA, and hooking together the amino acids to form a protein.

Inside the nucleus of the cell, even longer strands of deoxyribonucleic acid (DNA) reside. These are the master molecules of the cell and the functional elements of the chromosomes, because they provide the templates for all the RNAs that a given cell uses to code for the particular proteins that it produces. Organisms reproduce one another in their own likeness because offspring inherit the same set of DNA possessed by the parents. Different cells in the same organism have different shapes and different functions because a distinct subset of RNA molecules is produced from the DNA templates in different cells, thereby giving rise to a different set of proteins.

Well before the Second World War, cytologists (scientists who study the microscopic appearance of cells) had discovered a way of staining the DNA and RNA in cells. Thus, a more darkly stained cell could be presumed to contain more RNA (the amount of DNA ordinarily remains constant in all cells of an organism). Using this crude indicator of RNA quantity, a Swedish neurochemist, Holger Hydén, had found back in the '40s that nerve cells subjected to a high degree of stimulation stained more darkly for RNA. In 1959, Hydén published a scheme relating the composition of RNA in brain cells to the storage of memory. He postulated that the electrical disturbances caused by a unique pattern of neuronal activation would disrupt the sequence of chemical bases along the RNA chain. This, in turn, would change the code and produce a new and novel protein, which would activate a neurotransmitter at the synaptic ending of the cell—hence providing for a "Hebbian," or modifiable, synapse.

McConnell could not remember when he first became aware of Hydén's work on nucleic acids in brain cells, but thought it was during his travels in Europe about 1959. He had already become aware, however, of the importance of RNA through his contacts with developmental biologists who were looking to RNA as the repository of information used by half of a bisected planarian for regenerating the other half of its restored self. With the double-regeneration experiment having shown that learned information can persist through successive generations in some (presumably chemical) form, McConnell turned his attention to the possibility that RNA was the molecular repository of an animal's experience.

Since chemistry was not McConnell's forte, he persisted in trying to transfer the information 'bodily' from conditioned (trained) worms to untrained recip-ients. First he tried grafting parts of one worm onto another. Barbara Humphries, a graduate student in philosophy, went to work for McConnell just to earn some money, and became the most successful grafter of worms in history. Her success rate was only one in 20, but that significantly exceeded the highest success rate reported in the scientific literature at the time. Nonetheless, a rate of one in 20 successes was not enough to be cost-effective, so McConnell started trying to inject extracts of trained planaria into untrained recipients. This didn't work very well either. Getting a hypodermic needle inside a flatworm was about as easy as impaling a pancake with a telephone pole; and when by chance the needle did get inside a planarian, even the tiniest volume delivered from the syringe more often than not caused the injected worm to explode.

One day a letter arrived from another psychologist, J. Boyd Best, a friend of McConnell's. Best was studying the effects of changes in barometric pressure on the behavior of flatworms, and had discovered that cannibalism was a sensitive behavioral indicator of pressure. It occurred to McConnell right away that having one planarian eat another would be an excellent way of transferring the contents of a worm to its consumer. So trained worms were chopped into tiny pieces and fed to untrained recipients. As a control, untrained worms were also chopped up and fed to another group. When the two recipient groups were tested in the conditioning apparatus, those planaria which had eaten their educated counterparts responded nearly 12 times out of the first 25 trials, while the group that fed on untrained worm-chop responded an average of seven times in their first 25 trials. The difference was rather small but statistically significant.

McConnell's first version of these results appeared in the *Worm Runner's Digest*, rather than in a refereed journal. This was probably a mistake since the scientific community didn't know yet (and never did know) how seriously to take the contents of the *Digest*. It wasn't until 1962 that the results of the cannibalism experiment appeared in the *Journal of Neuropsychiatry*, and when it did it was scholarly and proper, with an erudite review of the history of experiments leading up to the improbable procedure being reported. But the popular press and media had by then spread the story, and most scientists first learned of it under the catchy headlines and humorous leads that the subject matter inevitably invited. While McConnell had made his view abundantly clear that traditional and valid science could be funny and fun as well, the levity of the situation took the edge off its credibility. To the scientific community in general, McConnell had hit upon something between an important discovery and a really good joke. The range of ambiguity was enormous. And McConnell couldn't resist adding to it. A few years later, in reporting the results of the cannibalism experiments at an International Congress of Psychology in Moscow, McConnell noted that his travel through the site of the 16th century Diet of Worms en route to the Congress had foreshadowed the work he was about to report—a jocular allusion that left his European audience as mystified as McConnell was amused.

Among scientists willing to examine the work seriously and in detail, there was an issue that went beyond that of public relations. Donald Jensen, a psychologist at the University of Indiana, from the beginning had challenged the notion that planaria could be trained at all. McConnell and Jensen had become locked in a traveling duel over the primary phenomenon—whether flatworms could ever be truly conditioned in the first place. It was a duel that Jensen lost over the long run, but the controversy it generated cast a cloud over all the planarian work, and provided another reason for cynics to question the validity of McConnell's more eccentric regeneration and cannibalism experiments.

It doesn't follow, however, that all who were skeptical were unaware of the potential significance of McConnell's research. Some in fact went to considerable effort to replicate and extend McConnell's findings, in the hope that the lowly flatworm could indeed shed light on the molecular mechanisms of learning. The lab that probably holds the record both for effort and skepticism was that of Melvin Calvin at the Berkeley campus of the University of California.

How Melvin Calvin, a Nobel laureate for his far-reaching discoveries on the mechanisms of photosynthesis, got mixed up in the worm-running controversy was an accident of circumstance. In 1961, McConnell was recruited by the Britannica Center for Studies of Learning in Palo Alto to become its Associate

Director and develop some programmed texts. He took with him to California Allan Jacobson, by now his graduate student, and Reeva Jacobson, who was in love with Dan Kimble and therefore anxious to be in the vicinity of Stanford to which Kimble had moved from Michigan. The Director of the Britannica Center was Alan Calvin, a cousin of the Nobel laureate at Berkeley. Through that connection, Melvin Calvin became aware of the planaria experiments, and invited McConnell to give a talk at Berkeley. McConnell's presentation impressed Calvin and the members of his lab enough to inspire the Berkeley group to try its hand at running worms. McConnell, not being a biochemist himself, welcomed the entrance of true biochemists into the field; and to spur them on, he gave them his equipment and arranged to have Allan and Reeva Jacobson work in their lab long enough to get them up to speed as worm runners.

By all accounts, the year at Berkeley did not go well. The Jacobsons felt that they were training the worms successfully, as they had done in Michigan. But Edward Bennett, the biochemist Calvin had designated to oversee the project, was unconvinced. Whether Bennett was unduly skeptical, or the worms really didn't work in California, would become a hotly debated and unresolved issue. Whichever was the case, for other twists in the story it is now necessary to delve into the career of Edward Bennett and his close colleague of many years, Mark Rosenzweig; because next to Hydén and McConnell, no one contributed more to research on the biological basis of memory in the early '60s than Bennett and Rosenzweig.

Lifting the Curtain on Brain Plasticity

Edward Bennett was born November 20, 1921, during the worst snowstorm of the decade, in Hood River, Oregon. Through his school years during the depression, he gravitated toward chemistry by default—he didn't care for languages or literature, was not proficient in math, nor was he drawn to physics, the glamour science of the '30s. He enrolled in the fall of 1939 at Reed College in Portland, attracted by the liberal educational philosophy of the institution that was regarded by many as radical at the time. As the war approached, he worked his way through college as quickly as possible, taking heavy loads and summer classes in order to graduate in 1942.

Linus Pauling, the future nobel laureate, was at that time recruiting the best chemistry students he could find to the California Institute of Technology, whose scientific brainpower was being redirected toward the effort to win the war. Enticed to Caltech by Pauling, Ed Bennett joined a research team in November of 1942 that soon relocated to a rural area of Florida north of Tampa for a year and a half of secret research on the chemistry of mustard gases. He contributed to the discovery of sulfide contamination in mustard gas, and published his first scientific paper as an outgrowth of this work. In 1945 he returned to Caltech, and finished his dissertation for the PhD in 1949.

The job market for scientists, who contributed significantly to winning the war, was tight in the post-war period, but Bennett was fortunate to obtain a position with Melvin Calvin at Berkeley. Calvin had pioneered the use of radioactive compounds to trace metabolic pathways, or the sequence of biochemical conversions in living cells, eventually winning the Nobel Prize for discovering the way in which carbon dioxide is converted into sugar by photosynthesis in plants. In 1949 his interests had turned to cancer drugs, and he needed a chemist to work out the metabolism of the anticancer drug azaguanine. Bennett spent a year manufacturing radioactive azaguanine, and another year following the metabolism of adenine, a vital cellular compound to which azaguanine is chemically related. (Adenine is a component of DNA and RNA.)

Azaguanine interferes with the metabolism of adenine, thereby stopping cells from multiplying.)

The Laboratory of Chemical Biodynamics was designed by Calvin to stimulate cross talk among the scientists who worked in it. Labs were laid out without walls in a circular array around equipment and conference space used in common. Calvin impressed those who worked with him that, while biology poses the most fascinating and important questions in science, chemistry provides the key to their answers. So he encouraged his chemists to think of ways to address the questions of biology, however complex or profound, by sharing insights, talking to one another across disciplines, and taking a shot at the seemingly unlikely or implausible approach that might lead to new ways of thinking. Ed Bennett found himself drawn into this ideal of the way science works at its creative best, when a meeting was arranged between him and two professors of psychology who were searching for chemical answers to the biological question of learning in 1952. Mark Rosenzweig, a new assistant professor recently arrived from Harvard, and David Krech, a former student of the famed learning theorist, Edward C. Tolman, had decided to look for chemical traces of experience in the brains of their laboratory rats.

Mark Rosenzweig had come to Berkeley as an indirect consequence of the loyalty oath controversy at the University of California in the early '50s. Tolman had refused to sign the oath and had exiled himself from California for a time till the controversy died down. He spent one year of his exile at Harvard. His colleagues at Berkeley had told him to interview prospective physiological psychologists while he was away, so this had brought him into contact with Rosenzweig.

Born in Rochester, New York in 1922 and graduated valedictorian of his high school class, Rosenzweig found that the field of psychology allowed him to combine his interests in both science and the humanities. He received a bachelor's degree and studied for a year in graduate school at the University of Rochester, then entered the Navy with the promise of a research appointment to the Naval Hospital in Bethesda. But when the Navy learned of his expertise in electronics, he was assigned to radar work instead and ended up on a seaplane tender in China.

After the war, Rosenzweig completed his graduate work at Harvard, where his research with Walter Rosenblith and Robert Galambos contributed important new information on the neurophysiology of hearing. Assured by Tolman that the loyalty oath flap in California was a transient aberration, Rosenzweig moved to Berkeley in the summer of 1951, prepared to pursue a career of research in auditory neurophysiology. While at Harvard, he had met David Krech, a psychologist exiled temporarily like Tolman from Berkeley because of the Loyalty Oath. Years of training rats to run mazes had convinced Krech that different animals develop different strategies for learning sequential tasks. Some rats, for instance, seemed to rely on visual cues for learning their way through a maze, while others appeared to formulate spatial images of the maze in learning how to solve it. Now both at Berkeley, Krech and Rosenzweig became good friends and colleagues in search of the substrates of learning.

As Rosenzweig was contemplating Krech's stimulating notions, he had begun to teach the newly derived consensus among neurophysiologists that neural transmission at the synapse depends on chemical rather than strictly electrical signaling. It had taken 40 years for the notion to become accepted that a nerve impulse causes the release of a chemical substance when the impulse reaches the nerve ending. Finally the evidence had become overwhelming that one such substance, acetylcholine, acts as a neurotransmitter, and that its action on the cell across the synaptic gap is terminated by an enzyme, acetylcholinesterase, that breaks down the acetylcholine once the next cell in the circuit has been activated.

Krech and Rosenzweig began to wonder if different neural abilities would be correlated with different chemical characteristics. Would an animal that uses the visual part of its brain more than other parts have higher levels of acetylcholine or of acetylcholinesterase in that part of the brain? Not being chemists, they called upon Calvin, whom Krech had met years earlier at Oxford. A meeting was arranged over lunch at the Faculty Club one day in 1953 between the psychologists and three of Calvin's chemists, including Edward Bennett.

Ed Bennett listened to the psychologists and thought to himself that the chances of detecting chemical differences in the brains of different rats or in different parts of the brain of one rat were not great. But given the potential payoff, he figured that it might be worth a few weeks of effort. To his surprise, the initial results were promising: rats that seemed to learn mazes by using visual cues had higher levels of cholinesterase in the part of the brain that deals with visual information, while rats that depended on spatial cues had higher levels of cholinesterase in other parts of the brain. The possibility of correlating brain chemistry and function suddenly seemed less remote.

At first, the newly formed research partners thought that different cholinesterase levels were genetically determined and fixed for a given rat. The more they examined their data, though, the more they began to wonder if the amount of cholinesterase in a rat's brain wasn't determined by the complexity of the learning task it had been given. Rosenzweig began to think about the summer of 1947 when Donald Hebb had taught a seminar as a visiting professor at Harvard, using as a text for the course a draft version of what would become his influential treatise, *The Organization of Behavior*, in 1949. Hebb described for his students how he would sometimes take a small group of rats home and make them family pets. After the more enriched experience of roaming about the furniture, playing with children, and exploring the complexities of a human habitat, the rats upon their return to the lab appeared to be smarter at learning things like mazes.

Rosenzweig, Krech, and Bennett decided to simulate Hebb's "free play" environment in the lab. Large enclosures were set up for housing rats in groups, containing objects for climbing and exploration, toys, mobiles, and other devices for providing a much more stimulating "environmental complexity and training" (ECT) environment than for the control rats housed in solitary cages.

Contrary to their expectations, the ECT rats were found consistently to have lower levels of cholinesterase in cortical regions of their brains than did the isolated controls. Enzyme levels, however, were measured by dividing the total amount of enzyme activity by the weight of the brain tissue from which the enzyme was extracted. This was done to compensate for the possibility of obtaining different amounts of starting tissue from different rats. It occurred to the researchers, therefore, that their measure of cholinesterase level (enzyme activity/tissue weight) might be lower, not because the numerator (enzyme activity) was lower but because the denominator (tissue weight) was higher. In rechecking their data, they discovered indeed that the weight of cortical tissue from ECT rats was inevitably greater than for the impoverished controls.

A neuroanatomist, Marion C. Diamond, was recruited into the group. Her careful anatomical analyses confirmed a number of small but reproducible morphological changes associated with environmental enrichment that seemed to account for the increased brain weights.

The results that began to flow out of the Berkeley lab were met with considerable skepticism. In the first place, the differences were quite small, on the order of just a few percent, and required statistical evaluation to demonstrate that they were real. Biochemists and anatomists in the 1950s, generally speaking, were not trained to rely on statistical methods for interpreting data, so were more prone to dismiss small differences as unreliable flukes. But time and again the measures came out the same.

The second reason for questioning the results from Berkeley was that they simply didn't correspond to the expectations of the day. It was generally assumed that the brain is static—that it doesn't change chemically or anatomically once it stops growing. For a group of researchers to claim that a few days of environmental enrichment would thicken the cortex and change the level of enzyme activity in the brain was a revolutionary notion. By dogged persistence and a steady stream of publications, the Berkeley group gradually got opinion shifting its way.

In time, more elegant and precise procedures confirmed their results and eclipsed their accomplishment. The apparent changes in cholinesterase activity did not, in the end, explain much about memory; and the anatomical changes turned out to reflect a variety of modifications studied more constructively at the microscopic level of individual nerve cells and their connections. The Berkeley group did, in fact, go on to more detailed studies of glial and neuronal numbers, sizes of nerve cell bodies and nuclei, dendritic branching and spine density, and the size of synaptic contacts. Though preliminary from the vantage point of today, it can clearly be seen that the work of Rosenzweig, Krech, Bennett, and Diamond lifted the curtain on the possibility that the brain is not so static after all; that the experience to which an animal is exposed leaves tangible traces which scientists with sufficient perseverance, rats, time, and technicians can uncover if they look long and hard enough. Their legacy became the understanding that learning leaves a tangible, detectable trace in the structure and chemistry of the brain.

"They're Loaded—They've Got Eyeballs and a Spinal Cord!"

Edward Bennett was not the only biochemist that Jim McConnell could have recruited to his cause. In many ways, Bernard Agranoff was a more likely candidate. Agranoff arrived at the University of Michigan in 1960, four years after McConnell had started the planarian experiments that by then were in full bloom. Already an accomplished neurochemist interested in the metabolic basis of information storage, Agranoff was caught up in the flatworm fever enough to try his hand at worm running. When the flatworm experiments failed, he turned to goldfish and hit a research gold mine. In that indirect way, his career was significantly influenced by McConnell, and he came to make important contributions to the study of the biological basis of memory, but not in a way that did McConnell any good.

Bernard Agranoff was born in Detroit in 1926 and raised in a lower middle-class neighborhood in the northwestern corner of the city. He showed an early interest in both science and art. By the time he was a teen-ager he had set up a chemistry lab in his basement, probably like the one by Eugene Roberts, the kid across the street who also had a basement lab and who would also later become a noted neurochemist.

Agranoff went to high school at Cass Tech, one of the nation's most noted public "magnet" schools for talented students. The first day of his class in Geometry for Art Students, the teacher was explaining that two triangles were similar if they had two sides and an angle in common. Agranoff asked her how she knew that, and she said "You feel it in your bones." That same day, he switched to a science math course. From art, he drifted into architecture, then finally to science, though by then he had missed the foreign languages required in the college preparatory track. World War II had begun, and a Navy V-12 Officer Training Program at the University of Michigan admitted him despite his deficiencies.

An Officer Candidate had five majors from which to voice a preference: pre-med, pre-dental, chaplain, civil engineering, or "deck" (ship's officer). With his interest in science, he leaned toward pre-med, which made his mother happy though he had no intention of practicing medicine. After two years at Michigan and with the war over, he began his medical studies at Wayne Medical School in Detroit, since he could start there a year earlier than at the University of Michigan. He didn't particularly like medical school, and was prepared to quit half way through, but his professors talked him out of it. In 1950, he received his M.D., then went to the Robert Packard Hospital in Sayre, Pennsylvania for a year of internship which allowed him to work part-time in a lab. Gordon Scott, the Dean of the Medical School at Wayne, recognized Agranoff's interest in research and steered him toward Francis O. Schmitt at M.I.T. Schmitt was a biophysicist before the term had been coined, using chemistry and physics to probe the ultrastructure of cells. When Agranoff saw all the fancy equipment that Schmitt had at his disposal, he was sold. Following his year in Pennsylvania, Agranoff took a postdoctoral position with Schmitt, a step ahead of the Navy recruiter who had not yet collected on the Navy's investment in Agranoff's education. Schmitt talked him in to working for a Ph.D., and he managed to pass his doctoral prelims, but the Navy could be denied no longer, so he was taken by the physician's draft and sent to Bethesda, Maryland. For a biomedical scientist, this was like being thrown into the briar patch. He never did complete the requirements for his Ph.D.

On the eve of his arrival at the Naval Hospital in Bethesda, Agranoff was told by a roommate at the bachelor officers' quarters to expect a refresher course in medicine and reassignment to sea duty in a few weeks. The next morning Agranoff found himself appointed as the teacher of the course, with his roommate one of the students. His roommate was at sea again in a few weeks, but Agranoff remained in Bethesda for two years. In addition to teaching, he was given the task of running the clinical blood lab—generally a low-pressure position except when a Congressman's blood sample got displaced or a specialized chemical procedure was called for, like the re-gilding of gold braid on an admiral's coat sleeves.

By the end of his two years at the Naval Hospital, Agranoff had begun to think seriously about the biochemistry of the nervous system. His mother's chronic depression may have been the reason. Watching her go in and out of the hospital many times, he couldn't help thinking that a better way to treat her condition should be found. Also, his sister was a psychiatric social worker, and one of his best friends had become a psychiatrist. In the mid '50s there was little research that could be called neurochemistry, but Agranoff's boss at the Naval Hospital, Roscoe Brady, had set up what would become a fine neurochemistry lab at the National Institutes of Health across the street, and he offered Agranoff a job. Jack Buchanan had in the meantime moved from the University of Pennsylvania to MIT, and invited Agranoff to return there to finish his doctoral dissertation, but the prospect of going into his own research lab immediately was more appealing, so he accepted the offer from NIH.

In Brady's lab, Agranoff gained experience and grew in stature. He began by studying the distribution of glucose in the brain, using sugar molecules "tagged" with radioactive labels. From that he progressed to the isolation of an enzyme, which gave him methodological skill and confidence. Then, in the course of investigating lipid metabolism in the brain, he accumulated evidence for the existence of a novel naturally occurring molecule. This compound, CDP-diglyceride, represented a key intermediate in the synthesis of a very important family of lipid compounds in the brain, the inositol lipids, according to Agranoff's prediction. The idea had come to him in a flash of insight in 1957. The paper was published the following year, and very soon the compound had been synthesized in another lab and shown to act in the way Agranoff had predicted, thus confirming its existence. It was his first major discovery, and with it he had, in his own words, earned his biochemist's "union card."

He read the proofs of his paper on this discovery on the boat to Europe in September of 1958. Having married the previous year, he embarked with his wife on an extended working honeymoon in Germany. Within two months, he had hit the big time again. Working with Fyodor Lynen (who received the Nobel Prize in 1964), he helped discover a key metabolic step in the synthesis of cholesterol. For the second time in two years he had made or participated in a discovery of lasting importance. For much of his remaining year in Europe, Agranoff traveled around with his wife at a more leisurely pace than he otherwise would have allowed himself without the good fortune of these important early discoveries.

In 1950, the drug chlorpromazine (Thorazine) was synthesized in France, and by the mid 1950s was being used clinically in the United States to treat mental illness. Because the drug has the ability to calm and relax a patient without diminishing basic reflexes or interfering with normal thinking and behavior, it is extremely effective in controlling psychotic states. The lithium salts were first used to control mania in 1949, and the antidepressant properties of a class of drugs used originally for the treatment of tuberculosis were discovered in the early fifties. These compounds collectively ushered in a revolution in the treatment of mental illness, converting an intractable disability into a manageable condition.

At about the time that Thorazine was introduced as an effective antipsychotic drug, the compound meprobamate (Equanil, Miltown) was discovered to have useful sedative and anti-anxiety properties as well. Its use as a tranquilizer became extremely popular in the 1950s, but unlike Thorazine and newer anti-anxiety drugs like Valium and Librium, it inhibits normal reflexes and produces drowsiness to a significant degree.

The use of animals in research, of course, was indispensable in developing and testing these drugs to the point where they could be safely administered to humans. For the most part, it was rodents, cats, dogs, and primates that were used for this purpose (though the ability of the drugs to tame certain wild animals was a key observation in discovering the tranquilizers). For a few biologists interested in non-human animal behavior, the drugs offered a new way to explore the neural mechanisms of species-specific behavior.

Among the researchers venturing into this new area of psychopharmacology was Ekhart Hess, a University of Chicago ethologist famous for his research on imprinting in birds.

An ethologist is a biologist who studies animal behavior, with an emphasis on the adaptive significance of behavior in the animal's natural environment. Imprinting is the tendency of newborn (or newly hatched) animals to become attached to the first significant stimulus that they experience. Normally this would be the sound, smell, or feel of the mother (hence its adaptive significance), though a beeping toy or talking human will serve equally well in the mother's absence. The phenomenon has been studied most extensively in those species of birds that are neurologically mature at hatching, ready to waddle off at an instant in tow to the first sound or sight they encounter. When meprobamate was introduced, Hess began to question whether the drug would block imprinting. If it did, and if the biochemical and physiological mechanisms of the action of the drug could be discerned, then a clue to the underlying mechanism of this elementary form of learning would be provided.

As one after another wonder drug for the treatment of mental illness was introduced in the '50s, Agranoff found himself asking how each of the agents affected the biochemistry of the brain. When he discovered that no one knew for sure, he decided to try to find out himself. His drug of choice was meprobamate, and he teamed up with a pharmacologist down the hall at the Public Health Service, Julius Axelrod, to study the metabolism of the drug in brain tissue. Through Axelrod he met many people working with meprobamate, including Ekhart Hess. Ducks at a bird sanctuary in Maryland not far from Bethesda served as subjects for some of the studies that Hess was conducting on imprinting. Agranoff went to the sanctuary and managed to get a group of hatchlings imprinted to himself. He brought them back to the research lab at the Public Health Service, where they were allowed in the more permissive atmosphere of animal husbandry at that time to follow him around, quacking from room to hallway, as much at home as any of the technicians or PhDs in the nation's premier health research institution. Ducks are messy by nature, though, and the animal caretaker tired of cleaning up after them, so eventually they had to go. But they were there long enough to become one of the minor legends at the Neurological Institute – the sight of those ducks in tow to their surrogate mother, the neurochemist curious to learn if a tranquilizer could mellow the metabolism of the brain to the point of blocking the imprinting of an animal's earliest, most critical experience. It was, in retrospect, the first serious thought that Agranoff had given to the scientific study of memory storage and retrieval in the brain.

It turned out that meprobamate's influence on imprinting was probably indirect and not as enlightening as Agranoff had hoped, so he lost interest in the project, but not in the overall issue of learning and memory. He began to read the literature and to acquaint himself with the names of the scientists established in the field. Ralph Gerard's catch phrase of the time, "not a twisted thought without a twisted molecule," reflected the impact of the recent discovery of the structure of DNA and its significance for genetic information. It was natural that brain scientists would begin to think along the same lines as the molecular geneticists, and Agranoff was among them, but at a speculative level only. In the lab, he stuck to the straight, quality but safe type of biochemical research with which he had first made his name. When it came time to move on from the Neurological Institute, he had the good fortune to be offered a position at his alma mater, the University of Michigan, where Ralph Gerard had become the elder statesman of speculation on biological mechanisms of memory, and where a young psychologist, Jim McConnell, had set up a lab for the study of learning in an organism with the simplest nervous system, except for jellyfish, in the animal kingdom.

Jim McConnell and Bernard Agranoff shared a ready wit and a joyful approach to their work, but in other respects were very different. McConnell was expansive, Agranoff restrained. McConnell was daring, Agranoff careful. McConnell was intuitive and subjective, Agranoff rigorous and objective. At the Mental Health Research Institute, they mixed like oil and water. Ralph Gerard's best efforts to team up Agranoff's skill and talent in biochemistry with McConnell's innovative use of flatworms for the study of behavior failed. But McConnell was in his heyday at that time: well publicized, provocative, and controversial. Though Agranoff couldn't bring himself to team up with that much controversy, neither could he let the potential of a great research idea go untried. So in his own lab, in his own way, he began a concerted but frustrating effort to become a worm-runner himself.

McConnell's subjective evaluation of the worms' reaction to shock bothered Agranoff. He designed and built an automated apparatus that he hoped would eliminate any unconscious bias the observer might introduce into the experiment. Jonathan Levine, a student hired by Agranoff to run the tedious trials, moved a cot into the lab and spent days and nights for weeks trying to pick up evidence that the flatworms could be conditioned. Nothing worked. Automation had deleted (or in McConnell's view, defeated) the behavior. McConnell would later claim that an automated apparatus cannot substitute for an experienced observer when it comes to detecting subtle behavioral responses of the type displayed by conditioned planaria. Agranoff, on the other hand, needed a behavior that was a little more robust and unambiguous. With nothing but negative results to show for weeks of effort, he finally sent his assistant out to get some guppies at the local pet store. Mired for weeks in frustration at the far end of the phylogenetic spectrum, Agranoff was ready for an organism with a little more familiarity. Of guppies, he noted with relief at the time, "They're loaded – they've got eyeballs and a spinal cord!" They also behaved nicely in a gross, quantifiable way, learning to avoid shock coupled to light in a small aquatic shuttle-box. Thus sold on fish as experimental animals, he next turned to work with Danio malabaricus, the giant Danio. They were trained to swim in a rotating tank at a certain rate, then were tested for their ability to repeat the

performance after a variety of treatments (such as prolonged cooling) designed
to modify their brain metabolism. When a freeze in Florida interrupted his
supply of subjects one day, he switched to goldfish, which would become for
him and for many others in subsequent years the white rat of non-mammalian
but vertebrate biological research.

Working with goldfish took Agranoff back to shuttle-boxes and into a
new body of literature, including some of the excellent early work by Jeff
Bitterman and his colleagues at Bryn Mawr College on shuttle-box learning in
that species. In the course of a business trip to Princeton (to consult with a
company concerning meprobamate), Agranoff decided to stop off in Philadel-
phia to visit Louis Flexner at Penn and Jeff Bitterman at Bryn Mawr. Flexner
had begun a series of experiments on memory and brain metabolism in mice that
interested Agranoff very much; they had an enlightening meeting and enjoyable
conversation. Agranoff's visit with Bitterman, however, changed the course of
his career.

Bitterman, unlike Agranoff, was more interested in the acquisition of
information (learning) than in its persistence (memory). He was working on the
properties of appetitive learning (behavior reinforced with food rewards) in
goldfish, using movement into and out of a shuttle-box as the basic behavioral
task. In the course of their conversation, Bitterman complained that the trouble
with shuttle-box behavior was that once the goldfish learned the task, they never
forgot it. To Agranoff, who had monumental difficulty in getting his fish to
remember how to swim properly in the rotating fish tank in Ann Arbor, the
revelation that goldfish had a hard time forgetting their shuttle-box behavior hit
like a bolt. He left Bitterman's office in a daze, and when he reached the train
station he called Bitterman back.

"Did I hear you right?"

Yes, Bitterman said, the goldfish learn their shuttle-box task for life.
Agranoff knew then that he had found a form of memory robust enough and
durable enough, in a brain big enough for the metabolic studies that he really
wanted to carry out. His days of worm running were a thing of the past. Instead,
he capitalized on the robust form of memory displayed by goldfish, soon
showing that intracranial injection of puromycin had no effect on acquisition of
the shock-avoidance task, but prevented the formation of long-term memory.
How animals learned was still unclear, but new protein clearly had to be made to
enable them to store permanently what they had learned.

"Molecular Biology?
I Don't Know What That Is."

On October 5, 1957, the Soviet Union launched the first artificial satellite into earth orbit, inaugurating the space age and with it, a golden age of expansion in science and science education. The United States, shocked suddenly into the realization that scientific and technological supremacy over the Communist world could no longer be taken for granted, dramatically increased funding for scientific research, development, and training. Overwhelming public support made becoming a scientist honorable, patriotic, and even glamorous. James V. McConnell and Bernard Agranoff were among the scientists at early stages of their careers during this auspicious period. Others, like Mark Rosenzweig and Ed Bennett were established and in position to ride the wave of public support and enthusiasm for science. Most fortunate of all, perhaps, were those in school or just entering the mainstream of science at this time. Not only was there swelling psychological and financial support for research, but monumental breakthroughs were generating tremendous intellectual excitement in several scientific fields.

Samuel Barondes was 23 and in his final year of medical school at Columbia University in October of 1957. The choice of medicine as a career had not been decided until his senior year as an undergraduate. In fact, a course in experimental psychology had stirred his interest in behavior so much that a major in psychology was his first choice. But his family in Brooklyn had other advice. His uncle told him, "Do whatever you want, but go to medical school. Nice Jewish boys go to medical school. You can always make a living—and then you can do whatever you like."

It was good advice, as it turned out, but it didn't dissuade him from his interest in behavior and the basic science of brain function. As an intern at the Brigham and Women's Hospital in Boston, his veer toward research was further promoted by two factors: the Brigham and Women's Hospital had a research requirement as part of its residency program, and as a recent medical school graduate, he was subject to the physician's military draft then in effect. Both requirements could be satisfied by working for a time at the National Institutes of Health in Bethesda, to which Barondes moved in 1960.

Barondes joined the Clinical Endocrinology branch, with the intention of working on the pituitary gland. This is a small bean-sized organ attached to the base of the brain. It secretes a number of the body's most critical hormones, including those that regulate growth, metabolic rate, and reproduction. Its secretions, in turn, are governed by other hormones produced in the brain. Thus the pituitary is the endocrine organ most intimately related, both physically and functionally, to the brain. With his dual interest in endocrinology and the nervous system, Barondes thus thought the pituitary gland the ideal object of study.

Gordon Tomkins, an endocrinologist with a background in biochemistry, was at N.I.H. at the time. He was only seven or eight years older than Barondes, but his gregariousness, intelligence, and excellence as a teacher had already boosted him to almost legendary status at the Institute. With their shared interest in endocrinology, Barondes and Tomkins struck up a friendship.

One day Tomkins asked Barondes, "Do you know anything about molecular biology?"

"No, I don't know what that is."

"I want to teach you about that. Come on into my office."

So Tomkins, the teacher, took Barondes, the recent medical school graduate, into his office and began to tell him about the wonders of molecular biology that by then were beginning to be unraveled at a quickening pace. He told him of the double helix of DNA, which carries the sequence of chemical bases that code for everything; of the single strands of RNA that carry copies of the code from the nucleus to the outer reaches of the cell where proteins are made by assembling their constituent amino acids in the sequence ordered by the RNA code; and of the possibility that all the secrets of life collectively boil down to the coordination and control of *which* proteins have their production turned on or off at what times. It was a brave new world in its entirety for Sam Barondes, and he was enthused.

"This is terrific!" he told Tomkins. "I want to work on this."

He could, Tomkins told him, but it would have to be with someone else, as Tomkins himself was headed for France on a sabbatical leave.

"But there's a guy down the hall who you should work with, who is just starting to do interesting work in this area." His name was Marshall Nirenberg. He had only one postdoctoral fellow—Heinrich Matthaei—working with him, so Barondes joined the team of Nirenberg and Matthaei just before the team became famous in every molecular biology lab in the world.

Nirenberg had decided to try to decipher the language of the genetic code. The alphabet was known, but not the vocabulary. It was known that RNA, the template molecule, consists of only four different types of bases (adenine, guanine, cytosine, and uracil)—four different 'letters.' It was also known that any protein, the product molecule (and the 'sentence' in this analogy), consists of a particular number and sequence of about twenty different types of amino acids—a vocabulary of 20 different 'words.' The protein-building machinery of the cell thus translates a particular sequence of four different types of bases in RNA into a given sequence of twenty different types of amino acids in the protein molecule. Clearly it took more than one RNA base to code for each amino acid; and presumably it took more than two, since the total number of two-letter combinations that could be derived from a 4-letter alphabet is only 16 (4 x 4 = 16). A sequence of three bases would be more than enough (4 x 4 x 4 = 64) to specify 20 amino acids, but no one knew the triplet code of bases that corresponded to any of the 20 amino acids.

Nirenberg's strategy was to use synthetic polynucleotides—long strands of artificial RNA molecules made of only one type of base—to direct protein synthesis in a cell-free system. That meant that the components of the cell responsible for manufacturing protein were mixed with the synthetic polynucleotides after separating away the cell's naturally occurring RNA. The first polynucleotide that Nirenberg and Matthaei used was made up entirely of uridine (U). When they added a complete mixture of amino acids to their cell-free system containing poly-U, only one of the amino acids—phenylalanine (Phe)—was incorporated into a protein chain. This meant that an RNA sequence of U-U-U-U-U-U-U-U-U yielded an amino acid sequence of Phe-Phe-Phe. They concluded, therefore, that the RNA base code for Phe must be U-U-U (or possibly more than three of them).

Barondes did not participate in this discovery, but he arrived in Nirenberg's lab just in time to witness it, and he took part in the next series of experiments, using other synthetic polynucleotides to decipher the code for the other amino acids. He demonstrated the catalytic (enzymatic) properties of poly-U and showed that RNA attaches to a special RNA-protein complex called the polysome in order to stretch out the messenger RNA strand so that it could serve as the template for ordering the sequence of the amino acids into proteins.

The significance of Nirenberg's work was appreciated immediately, and within eight years was recognized by the award of a Nobel Prize. For Barondes, "It was an incredible experience. . . . I went from knowing nothing to suddenly being in one of the hottest labs in the world. It just sort of happened in a three month period before my very eyes." The good fortune of this first positive experience in scientific research affected him profoundly. He saw that basic science could really answer questions and solve problems at the core of the mystery of life.

His association with Nirenberg opened the door to opportunities throughout the scientific world. He could have capitalized on the flush of this first success and pursued the research gold mine that he had hit upon almost by accident. But his ultimate interest remained in the brain and endocrine system, and he still wanted above all to work with Tomkins, so he went to France for the final months of Tomkins' sabbatical there to renew the scientific collaboration that had begun in casual conversations in the halls at N.I.H. By this time, though, Barondes knew something about molecular biology, and wanted to turn the power of the new science in a new direction. Having seen what molecular biology could do for the mystery of life, he had no reason to assume that it would be less successful in unraveling the mysteries of the brain.

Like McConnell, and virtually every neuroscientist interested in mechanisms of memory since 1949, Barondes was influenced by Hebb's proposition that learning involves a growth process or physiological change at the neuronal synapse. But unlike McConnell, Barondes had the biochemical knowledge to formulate a precise experimental question that tested one implication of the Hebbian synapse in terms of the emerging science of molecular biology. If learning requires a growth process at the synapse, it presumably needs new protein for that growth; and that, in turn, might require the production of new RNA to provide the code for production of the appropriate protein. If the synthesis of new RNA could be blocked, then the production of the new protein required by a Hebbian synapse might be blocked, and learning would not be possible.

Back home at N.I.H., Barondes teamed up with Murray Jarvik, another friend of Tomkins, to experimentally test the notion that memories could be blocked at their molecular source. They used the drug actinomycin-D, an antibiotic known for its ability to block the formation of new RNA. Sizable doses of the drug were injected into each side of the brains of mice. Four hours later, the mice were trained to avoid an electric shock by staying out of one area of their training box, then tested one to three hours later on their ability to remember how to avoid being shocked.

The results of this experiment were disappointing. In the first place, the drug made the animals very sick, so the persistence of memory could be tested only over a matter of hours. Secondly, the mice managed to retain their memory of how to avoid the electric shock, despite their acute, fatal illness. Finally, the drug managed to inhibit most (83% on average) but not all RNA synthesis; so the residual amount of RNA that the brain managed to produce may have been enough to account for the memory storage which apparently occurred. In short, the outcome was ambiguous and inconclusive.

Barondes and Jarvik were not the first to try to block learning by interfering with the macromolecular machinery of the cell. A few years earlier Wesley Dingman and Michael Sporn had impaired the ability of rats to learn a swimming maze by injecting 8-azaguanine into the brain. This drug was thought to interfere with RNA production, but the precise biochemical effects of the drug were not known. Also, in 1962, the noted anatomist at the University of Pennsylvania, Louis Flexner, inaugurated a series of experiments based on the blockage of protein synthesis directly. He used the drug puromycin, which one of his co-workers, Gabriel de la Haba, had previously shown to be capable of profoundly inhibiting protein synthesis.

Flexner's first attempt to block memory with puromycin was disappointing. Like Barondes, he achieved about an 80% inhibition of new protein synthesis, without impairing the ability of mice to learn or remember how to choose the correct arm of a Y-maze to enter in order to avoid an electric shock. Flexner discovered, however, that injecting the drug at multiple sites in the brain eventually could block memory, but with a complicating twist. To block recently acquired memory, injections into only a restricted region of the brain were required. But to block longer-term memories, the drug had to be dispersed more widely through the brain. It seemed that the changes associated with storage of information were "spreading out" over wider areas of the brain as time went on.

At the annual meeting of the American Psychological Association in September 1964, Barondes gave a lecture that tried to summarize the state of understanding about molecular mechanisms of memory at that time. From his training in psychology, he had derived a firm commitment to the Hebbian notion of modifiable synapses as the cellular basis for learning. But from his experience in molecular biology, he was able to go beyond the thinking of McConnell, Rosenzweig, Krech, and Bennett by pinpointing a number of specific, potential molecular candidates for the role of synaptic modifier. He pointed out that chemical changes might be functional (increased synthesis or release of neurotransmitter) or structural (construction of new axonal or dendritic membrane). He noted that synaptic changes might be only one of several processes necessary for storing memory, and they might be involved only at later stages. Finally, he pointed out with clarity for the first time the distinction between molecules as *embodiments* of the stored information, and molecules as agents of functional or structural changes occurring in a neural matrix to make memory possible.

The lecture was published as an article in *Nature* on January 2, 1965. Some of its basic insights can still be detected in the thinking of memory researchers today. But that fundamental tenet—that molecules are agents of a process for storing memory rather than repositories for the memories themselves—was to be challenged to the core by the time the year was out. McConnell with his flatworms had set in motion experiments that would turn 1965 into a year of revolution in the scientific study of learning and memory.

Acrobatic Rats and a Call to Kansas

On August 21, 1965, I was graduated from Texas Tech University with a major in chemistry and a minor in zoology. I sat in the balcony of the auditorium that hot Texas night, happy to be finished with four arduous years but feeling detached from the ceremony below me, my mind already elsewhere. I was by then aware of revolutionary developments in the field of learning and memory, and was anxious to move on to graduate school where I could acquire the skills to unravel the biochemical basis of brain function.

That such a thing was possible erupted into my awareness one morning during the fall of my sophomore year at Texas Tech, when I heard a brief news item on the radio that made me catch my breath. A Swedish scientist, according to the report, had just succeeded in detecting changes in the composition of nucleic acids in the brains of rats taught a wire-climbing trick. I knew just enough chemistry at that point to realize that nucleic acids were important; and I had developed a blind-faith notion that brain function is driven by the chemical codes and consequences of the molecules in its cells. I called the radio station immediately for a copy of the story from the newswire the announcer had read. It arrived the next day, and I wrote immediately to the scientist named in the story, Holger Hydén in Goteborg, Sweden.

When the reprints from Sweden arrived, they unveiled a fascinating probe into the chemistry of brain function. Hydén had succeeded in measuring changes in the concentration of RNA, and in the base composition of the RNA, in tiny amounts of brain tissue of rats trained to climb a narrow wire to reach a food reward. This difficult motor learning task involved the vestibular nucleus, the region of the brain from which the cells for RNA analysis were obtained. These changes in RNA meant that the chemical composition of the nerve cells of the vestibular nucleus was changed by the animal's experience in learning to climb the wire.

The following semester, I took a psychology course taught as a lively and broad but objective science by Pascal Strong. One of the requirements of the course was a term paper, and it was that assignment that gave me an opportunity to bring to bear for the first time my fledgling knowledge of research on brain mechanisms of learning and memory. At that point in time, my knowledge consisted mainly of the early work of Bennett and his colleagues at Berkeley on changes in cholinesterase levels in rats from enriched environments, and in the work of Hydén on RNA changes associated with learning. But I had also heard of the curious cannibalism experiment with flatworms at Michigan (who hadn't?), and of other hints that the nucleic acids so vital for the storage of hereditary information might also be involved in the storage of experiential information as well.

Work on the term paper was a labor of love from the beginning. The more I read, the more I realized that there were legitimate, established scientists who thought like I did about the brain, behavior, and its underlying chemical mechanisms. Even more tantalizing was my sense that I had about as much insight as any of them into what might be going on at the molecular level. In retrospect, this was testimony to our shared ignorance more than my insight, but at the time it was heady reinforcement of the career choice I had made.

The paper gave me my first opportunity to evaluate and synthesize various notions about how memory might be stored in the brain: The flatworm cannibalism experiment showed that memory is stored in a chemical form. Hydén's experiment showed that the chemical form could change in concert with the animal's experience. And the studies at Berkeley showed that chemical changes at the synapse might be the means by which the alterations in neural circuitry underlying memory were accomplished. Thus, I wrote in May of 1963:

A novel experience is transmitted to the central nervous system as a new impulse frequency. The molecular changes specified by Hydén occur; a new substance is synthesized which activates acetylcholine more readily and the impulse crosses the synapse. The new substances continue to be synthesized so that synapse transmittance is facilitated the next time the new impulse arrives.

There it was: my first theory of the mechanism of memory. What was meant by impulse frequency, what the substances were that activated acetylcholine, or how they continued to be synthesized were issues for another day. The idea that memory involves a change in synaptic efficacy was hardly original; in fact, I cited Hebb extensively in my paper. But at the time it was written, there weren't many scientists saying, as I was, that McConnell, Hydén, Rosenzweig, Krech, Bennett and Diamond were all looking at different pieces of the same puzzle.

Later on, as I developed skepticism toward the notion of the Hebbian synapse, my theory lost its appeal; and in time I came to doubt the scheme in its entirety. But I made an A+ on the paper, and that reinforced my feeling that the gap between memories and molecules could be spanned. In the Spring of 1963, I believed that I had as good a chance as anyone of building that bridge.

The National Science Foundation (NSF) over the years has funded a marvelous program that enables undergraduate students to become involved in research early in their academic careers. The Foundation provides a small stipend for students to work with active researchers on specific projects. I knew about it because Texas Tech had applied for one; but I didn't want to stay in Lubbock that summer, so I wrote to NSF for their list of institutions where the programs were being carried out that year. The two institutions and topics that sounded closest to my interests were at the University of Chicago and the University of Kansas. Both responded to my initial inquiries, though I was particularly impressed that the chairman of the department at Kansas, Dr. Frederick E. Samson, had written a personal letter responding to my specific questions. I applied for summer fellowships at both Chicago and Kansas. Most institutions strongly favored their own students. Thus I was not surprised to be turned down at Chicago, but was elated when Kansas decided to take a chance. The letter accepting me as an NSF Undergraduate Research Fellow at the University of Kansas under the direction of Dr. Frederick E. Samson arrived on April 25, 1964, turning the end of my arduous junior year into a celebration and brightening my prospects for the future. With finals mercifully out of the way, I boarded a bus the first evening in June for the 18-hour ride to Lawrence and a fateful summer of work in the lab of Fred Samson.

Samson

"Come See Me in Chicago"

Nothing in his family history provided a foretaste of the interesting life that Fred Samson was to lead. His father, grandfather, and great-grandfather had all been firemen in and around Boston, where the family had lived since Kindred Samson arrived at Plymouth on the *Mayflower* a day or two before Christmas in 1620. His parents retained the puritanical demeanor of their heritage, wearing plain clothes, eating plain food, leading plain lives. The younger Samson and his four older sisters were raised in a household without alcohol or other obvious vices, and without the energy and emotion of the immigrant Italian and Armenian families around them. Scholastically, the Samson children felt little pressure or encouragement from home to excel, but Fred Jr. managed to make a B average and do well in math (geometry was his favorite subject), somewhat to his surprise since he didn't think he was particularly smart. Upon his graduation from Medford High School in 1936, Fred Samson decided to go for a career in osteopathy for no apparent reason other than his parents' sense that it was "the kind of thing I ought to study."

In those days the practice of osteopathy was based on a narrow focus on the skeletal system, with little basic science orientation. Samson soon recognized that his instructors knew very little about the human body or disease. He came to hate his classes and dread the clinical sessions where he had to deal with patients. Most of the patients had serious medical problems, and had come to osteopathy as a last hope. Samson knew that osteopathy as it was then being practiced couldn't help them, so he spent his time talking to the patients rather than "treating" them, stretching out each visit as long as possible. By this tactic, he managed to see fewer patients, and avoided to some extent giving those he did see the false hope of a cure; but his instructors were exasperated by his lack of productivity.

To divert his mind and body from the ordeal of osteopathy school, he pursued a long-held interest in acrobatics. He dropped into a gym on Massachusetts Avenue in Cambridge one day, and, seeing the ropes and cables used in training gymnasts, knew that tumbling was something he had to pursue. He befriended the owner and soon was a student of acrobatics, paying for his lessons by helping the teacher train others. He apparently had a talent for backflips and head stands, because his teacher told him one day that he should go into show business. They developed a skit, with Samson dressed as a bellhop running through an acrobatic routine culminating in an impressive hand balancing act. It was good enough to get him bookings in nightclubs, burlesque houses, and the theaters that were still providing live entertainment between movies. By the time he had finished osteopathy school in 1940, Fred Gray, as he had renamed himself for the stage, could get bookings from New York to Vermont. At three dollars a night, the pay wasn't great, but anything was better than the practice of osteopathy.

War loomed on the horizon. Fred Gray, acrobat extraordinaire, got his draft notice in March, 1941. He thought about the show business world into which he had fallen. He was good at what he did, but he was also realistic. Unlike the starstruck and often aging performers with whom he shared a nightly billing, he recognized that there was little hope of becoming famous enough to command more than a series of low-paying one night stands in out-of-the-way places into the indefinite future. With nothing more than that to look forward to, he decided to let himself be drafted.

Though it was illegal for an enlisted man to have a car at Camp Edwards on Cape Cod, Samson had one and the officers tolerated it because they could get him to drive them into Boston for nights on the town. Furthermore, since Samson had never learned to drink during his straight-laced upbringing, he provided them with a sober driver on such occasions. So it was that on Sunday afternoon of December 7, 1941, Samson and a quartet of officers were AWOL in Boston when the news that Pearl Harbor had been attacked came over the radio. In a panic, Samson raced around town to gather up his dispersed, reveling riders. One officer, found drunk and naked in an apartment in the Back Bay, refused to come, but the other three were gathered up and returned to the base, past MPs who were more concerned about getting their disheveled forces assembled than on checking whether they had been absent with or without leave.

Samson's unit was shipped out to Australia, where the Americans were received with gratitude since that island nation was virtually defenseless -- its entire armed forces being tied up in Europe and Africa. Then the unit was dispatched to New Caledonia, which was nominally in hostile (Vichy French) hands, but where the Americans actually received a tentative welcome from the native inhabitants. In anticipation of a Japanese invasion, the American troops were dispersed throughout the island into small guerrilla units. For nine months, they waited. With endless volleyball games on the beach, swimming in the mountain streams, and ample opportunity to hone his acrobatic skills, Samson enjoyed "the best camping trip I ever had." He also tasted spiced food for the first time, and began to grapple with the world of ideas. Dark came by seven o'clock in the evening, and the camps were blacked out for security reasons. There was nothing to do but talk; so Samson and his fellow soldiers, drawn as they were from every part of the country, various traditions, and different religions, would talk at length about anything and everything. Samson realized that he was just as provincial as the kid from the farm. Night after night, he wrestled with a new set of notions and ideas—sometimes listening, sometimes arguing—with the growing awareness that life is not at all as plain as his monotonic upbringing in Medford had been.

In a more immediate sense, his life became a lot more interesting when his unit ended its nine-month camping trip on New Caledonia and was shipped to Guadalcanal. On August 7, 1942, the Marines had invaded Guadalcanal in the first major counteroffensive of the war in the Pacific, and managed to capture portions of the island, including the airport, from the Japanese. When Samson's unit arrived, the surviving marines were still in a daze. Fighting had been fierce, and it continued in the interior of the island. Samson was a medic, given his training in osteopathy, but at that time being a medic amounted to barely more that litter-bearing. The part of the island held by the Americans was under frequent attack from Japanese aircraft, so Samson spent a lot of time in the trenches. Watching dogfights in the sky, dodging bombs and bullets on the ground, and doing what little he could for wounded marines, Samson experienced the grim reality of a war far different on Guadalcanal from what it had been like on New Caledonia.

As in all wars, there were long interludes of boredom among the occasional moments of high drama. Samson decided during one of the boring interludes, that he would never be in a more appropriate setting for dipping into Tolstoy. Primed by the nighttime philosophy to which he had been exposed on New Caledonia, he found that he enjoyed literature; so his threshold was lowered when a pamphlet entitled "University of Chicago Home Study" caught his eye. "Through the U.S. Army, you can enroll in home study through the University of Chicago," the text said. He filled it out and sent it in, selecting English Literature 1 as his home study course. He had taken nothing of the sort at the college of osteopathy.

Mrs. Baskerville, a tutor assigned to English Literature 1 in the Home Study Program back in Chicago, couldn't believe it when she got an application from a private on Guadalcanal. All she knew from the news was that a heated battle for the island was underway. She immediately dispensed an English Literature textbook to the war zone; and after she saw over the ensuing months that Samson was a pretty good student, wrote to him saying, "If you come out of this war alive, come see me in Chicago."

Samson was about to get out of the war alive. As a medic, he had experience dealing with malaria and working in malaria control projects on Guadalcanal, so the Army decided that he should return to the United States to help train the thousands of troops that were being sent into the South Pacific in preparation for the anticipated invasion of Japan. He ended up at Fort Ord in California, working in a dispensary. Since the invasion of Japan was expected to cost a million American lives or more, the bombing of Hiroshima and Nagasaki, and the truce that quickly ensued, came as a welcome relief. Samson, having served overseas and in combat, was among the first to be discharged. Going back into show business after the profound experiences of the war was out of the question. He decided instead to go to Chicago and look up Mrs. Baskerville.

In ordinary times, under ordinary circumstances, Fred Samson would never have been admitted to the University of Chicago. His high school grades were far too mediocre, and his training at the osteopathic school counted for little. But in the aftermath of the war, other factors did count. There was, first of all, a sense of obligation on the part of the faculty toward those who had fought the war; and there was secondly Mrs. Baskerville, who went out of her way to get the returning veterans who had shown some ability in their home study courses enrolled in the University. Samson had taken the entrance exam and scored reasonably well, but not at the level demanded by most of the graduate programs. Thus, while the University was inclined to admit him, there was no single discipline for which he could claim a particular aptitude, and no department especially anxious to have him.

Overhearing the conversation relating to this dilemma was an advisor from the Department of Physiology. Since Samson had scored well on the quantitative aspects of the exam, the advisor suggested that he might persuade Physiology to take a chance on him. Thus, Samson was admitted to graduate study at the University of Chicago in the Department of Physiology. But he lacked the elementary background necessary for graduate study in the biological sciences, so he spent his first two years taking basic courses in math, physics, and chemistry. He did well in those subjects, vindicating readily the University's gamble on his potential. In his course on Boolean algebra, for example, he got the highest grade in a class full of math majors. He might have pursued mathematics or the physical sciences, were it not for an anxiety about math that he never overcame. While he did well in the subject, he didn't know why. He could solve the problems, but he didn't understand how. It was sometimes akin to a mystical experience—he would ponder a problem for an hour or two, unaware of the passage of time, and awake from a trance-like state with the solution. Perhaps it was a precursor of the fluid (some would say chaotic) way his mind seemed to work in the presence of his students and colleagues in the years to follow.

Notwithstanding his facility in math and the physical sciences, he found physiology to be equally enjoyable and rewarding; so he stayed in the Department that had given him a chance. With basic courses out of the way, he took up the curriculum in physiology and excelled to the point of becoming one of the best students in the department. The University of Chicago was at that time near the peak of its influence in the biological sciences (and many other fields). Ralph Gerard, already an elder statesman in neurophysiology, was giving brilliant lectures at the time; and Julian Tobias was publishing his provocative concepts of nerve excitation a decade before they were appreciated. Samson took it all in—with trepidation at first (he was so intimidated by Gerard's reputation that he was afraid he would flunk out the first time he attended a lecture)—but with growing confidence backed up eventually by impressive academic achievement.

Carl Sagan was a student at the University of Chicago at the time, as was

Horace Freeland Judson, the noted historian of science. A fellow student of Samson's in physiology, Gene Aserinsky, confided one day that in his studies of sleeping infants, he noticed that their eyes would dart around beneath their near-transparent eyelids in bursts of activity. The two graduate students pondered the mystery of this observation, unaware that in the hands of Kleitman and Dement, it would later lead to the discovery of rapid eye movement sleep and the dream cycle. "You'll never graduate with this," Samson advised his friend.

It appears that Samson got better advice than he gave. Daniel Harris, a cell biologist in the department, urged him to turn his interest in math toward the study of metabolic rates in various types of cells, ranging from yeast to muscle. At that time, the Department of Physiology at the University of Chicago was uniquely innovative in its biophysical and mathematical orientation. Francis O. Schmitt, a pioneer of the American school of biophysics with his electron-microscopical studies of connective tissue collagen, was awarded an honorary degree by the University while Samson was there. Zim Herron gave a course in kinetics (as applied to physiology, rates of metabolic reactions), and Samson was the best student in the class. The mathematical approach of kinetics to physiological phenomena appealed to him, so Herron became his formal advisor. But Daniel Harris was the faculty member that befriended him, provided him with laboratory space, and acted as his advisor *de facto*. Harris and Samson worked out a protocol of experiments testing the effect of various-sized fatty acids on metabolic rates in different cell types, seeking to find if different tissues metabolize fat in different ways. The work was successful enough to yield a doctoral dissertation, which Samson completed in 1952.

The chairman of the Department of Physiology informed him as he completed his dissertation that there was a job opening at the University of Kansas in Lawrence. "Where's that?" Samson asked, knowing nothing of the plains beyond the outskirts of Chicago. The chairman described a small and pleasant college community in the foothills of the Ozarks near the eastern edge of Kansas, just 40 miles west of Kansas City. It was a nice place, he said, and besides, the present and former chairmen were both from the University of Chicago, "and they like University of Chicago graduates." Kenneth Jochim was the Chicago alumnus then chairman of Physiology at the University of Kansas. While the medical school was located in Kansas City, medical students spent their first two years, and took their basic physiology course, on the Lawrence campus. Jochim liked Samson, as predicted, so with no teaching experience to speak of, and a rather unusual background for a physiologist, Samson became an Assistant Professor of Physiology at the University of Kansas in Lawrence in the fall of 1952.

"You Want to Tell Yourself What to Do"

When Fred Samson became a faculty member at the University of Kansas, he set up a lab to continue his research on metabolic rates in different types of cells, with an emphasis on yeast. In the summer of 1953, the chairman asked Samson if he could use a technician, since a coed interested in microbiology had asked if there might be work that she could do in a lab. Samson said "Sure", thus beginning a long and productive collaboration with Nancy Dahl and her husband, Dennis.

Nancy and Dennis met at a dorm mixer (informal dance) at the University of Kansas in the fall of 1950. Dennis was one of four children born to German Mennonite farmers driven from Kansas during the dust bowl days of the '30s to Oregon and Washington. When the winds died down and the rains returned to the plains, the family moved back to Wichita, then to the small town of Colby in northwestern Kansas. Only one of Dennis' siblings had reached high school, and none had finished, so his goal growing up had simply been to become a high school graduate. Encouraged by a teacher during his senior year to consider college, and bolstered by the highest score in Kansas on a placement exam in chemistry, Dennis enrolled at the University in Lawrence, which he had never seen.

Nancy's grandmother was a pioneer, widowed at the age of 22 on her western Kansas homestead. Nancy's mother was married to a County Agent—one of the legions of rural consultants that helped American agriculture survive the depression and prosper in its aftermath. But he left his family in Colby when Nancy was 15. Nancy was dyslexic, and didn't do well in school, but like her mother and grandmother she was determined to survive. "At some level you know you're not as dumb as they're saying you are," she reflected, cultivating an interest in music, art, and science. She first came to Lawrence to music camp the summer before her junior year in high school. Her original plan, as mapped out in a fourth grade essay, had been to marry a doctor and get him to let her work in his lab. With maturity came the realization that she could work in her own lab, but medical technology was the only lab profession for a woman of which she was aware, so bacteriology became her major.

Dennis was one of 49 graduates from Colby High School in 1949, and Nancy was one of 50 from the class of '50, but they didn't know one another until that night at the mixer in Lawrence when one of Dennis' friends pointed to Nancy and said, "Hey Dahl; here's a girl from your home town." Nancy, who was dancing with someone else, turned to look at Dennis just as someone took a picture. Dennis would carry in his wallet ever after that photographic record of the moment they met.

Neither of them had any money, so their courtship centered around free events at the university, and Saturday night walks around Colby the following summer, where they both had returned home to work. With the growing military conflict in Korea, Dennis had to face the prospect of induction into the Army at the same time that his relationship with Nancy was growing more serious. On January 23, 1952, Nancy and Dennis were married in Danforth Chapel, at the crest of the hill on the campus of the University of Kansas. With a borrowed car, they drove to Topeka to see an Abbott and Costello movie, found a motel room for a couple of dollars, and returned from their honeymoon the following morning, as the car had to be back to its owner by 9:00. They were fortunate to find a housing co-op where they could sleep and eat for $75 a month. Before the semester was out, Dennis was in the Army.

Every dollar that Dennis sent home was matched by the Army, so he kept a few dollars a month for himself and sent the rest to Nancy. She was into her second year in college after a shaky start, still pursuing microbiology in search of some kind of career in science. By the end of her third year in Lawrence she had matured, had come to grips with the departure of her father, and had improved her academic record. It was time to get a job in a real lab, she felt, so she began to scour the science departments of the university in search of an opportunity. Dr. Jochim, the chairman of the Physiology Department, offered her a job cleaning animal cages, and she did that with time to spare, so he asked her if she would be interested in working for the new physiologist in the Department, Dr. Fred Samson.

Fred Samson and Nancy Dahl hit it off well. Sharing family backgrounds distant from academia, and exuding an enthusiasm for science in its near pure form, the vivacious girl from western Kansas and the extroverted assistant professor were the perfect match of personality and circumstance. Science was still a gee-whiz phenomenon for Samson; he believed it was something to enjoy and to have fun with. The labor of the lab was as much recreation as vocation. And this played perfectly into Nancy's expectations. So the two of them formed an energetic team—one of only two or three that were doing any research in the Physiology Department at the time.

The second day that Nancy worked for him, Samson sat her down and told her, "You don't want to be a microbiologist. You don't want to be a lab technician. You want to tell yourself what to do. You don't want people telling *you* what to do." It was typical of him to proselytize his own experience, to get others to see that where you came from didn't have to limit where you could go.

So Nancy Dahl, who a week earlier didn't know how to spell "physiologist," determined to become one.

After he had taught for some time, Samson was asked what kind of teacher he was. "Actually, I'm an inspirational teacher," he replied. "I walk in there, and whatever I have an inspiration about, I teach about."

In reality, his teaching on the whole seems not to have been as haphazard as he suggested. Dennis and Nancy, who would eventually take a number of courses from him over the years, rated him a good teacher. It is true, Dennis reflected, that his lectures were occasionally of a purely "inspirational" nature. "When he was really busy with doing other things, he didn't prepare very well, and he did just lecture off the top of his head. But when he was prepared, he was always well organized, and [his lectures] always told a story."

It was apparently in the lab that Samson's ability as a teacher found its highest expression. By all accounts, he had a remarkable talent for picking up on the things that students did right, and criticizing the things they did wrong without puncturing their egos. "When you made a mistake," Nancy explained, "he always had a way of never, never putting you down. He would say, 'Yea, well, that's almost right, but I think if you would look at it this way' or 'I think if you would do this . . .' "

When Dennis returned home from the Army in the spring of 1954, Nancy infected him with enthusiasm for the work that she and Samson were doing. The program of federal grants from NIH and NSF was still in its early stages, and grants weren't hard to obtain. Samson got a grant, and hired Dennis to join the team, as the emphasis of his research began to shift from yeast to rats.

In converting to work on animals, Samson was influenced by the feeling that the work with yeast had about run its course, and, perhaps more importantly, by running across a little book by Harold Himwich entitled *Brain Metabolism and Cerebral Disorders*. Himwich was a pioneer in the field of neurochemistry, working at a state lab in Galesburg, Illinois. His book, published in 1951, represented one of the first lucid arguments that the nervous system is amenable to biochemical as well as anatomical dissection, and that a science of the chemistry of nervous tissue might be a realistic possibility. Samson realized that the same measurements he was making on the metabolism of yeast cells could be made on the metabolism of brain cells. Plus research on the nervous system had the added attraction of touching on that most exciting and profound of all mysteries: the workings of the brain. There was no reason that he and Nancy and, now, Dennis—with their considerable energy, un-bounded enthusiasm, and the certain satisfaction that hard work in a noble cause brings— couldn't embark on this grandest of scientific adventures.

"You Must be Doing Something Wrong"

To smooth the transition from microbes to mammals, Samson stuck with his focus on metabolism. One of the first projects that he and Nancy undertook was the measurement of blood sugar levels in diabetic rats. Nancy had to do the sugar assays, which involved the tedious transfer of an endless number of liquid samples from one vial or tube to another by pipeting (sucking them up in a glass tube). She remembers one particular day in the spring of 1954, after pipeting hundreds of samples from dawn to dusk, she tried to drink a milk shake that night and literally could not suck it up into the straw!

With Dennis on the team, they decided to pursue the possibility of a relationship between fatty acids (compounds found in vegetable oils and animal fat) and diabetic coma. From his previous work on yeast and muscle cells *in vitro*, Samson knew that the fatty acids could inhibit cellular metabolism. How would the fatty acids affect an intact animal, especially a diabetic animal already suffering from the consequences of insufficient glucose? When they injected fatty acids into rats to measure their effects on blood glucose, they found to their surprise that these compounds induced unconsciousness in control animals. The fatty acids were inducing sleep like a narcotic! It was reversible, dose-dependent, and varied with the size of the fatty acid molecule. It was their first major discovery as a team, and it reinforced their belief that a scientific relationship between the chemistry of the brain and behavior could indeed be pursued.

Another experiment inspired by work that Himwich had done had to do with survival time of rats under anoxic (lack of oxygen) conditions, the loss of consciousness due to anoxia being another form of coma. Rats of different ages were exposed to pure nitrogen atmospheres, to determine how long they could survive without oxygen. All mammals are highly dependent on oxygen, and can't survive without it ordinarily for more than a few minutes at most. Thus it came as no surprise when Nancy conducted her first experiment and found that adult rats placed in an oxygen-free chamber of nitrogen gas died within the predicted two or three minutes; but she was shocked to discover that newborn rats could go many minutes longer than the adults before succumbing to the lack of oxygen! Obviously, for rats the dependency on oxygen was not so critical at birth; but it became increasingly vital during the ensuing weeks of life. The first time she saw the baby rats still squirming for minutes after the older rats had died, she couldn't contain herself. Samson was in the middle of a lecture, but she knocked on the classroom door and frantically motioned him out into the hall to tell him the news! Such was the level of their excitement in the early days of their work together.

Samson felt that they were beginning to make inroads into the chemistry of brain function, and decided to ask Himwich—by then the elder statesman of American neurochemistry—if he would be willing to take on some visiting scientists for a summer. Impressed with the Samson and Dahl experiments on developmental changes in survival time under anoxia, one of his major interests, Himwich offered a three-month appointment to Fred Samson and to Nancy and Dennis Dahl for the summer of 1957. Just that year, Georges Ungar, a French scientist well known for his work on tissue response to injury and a pioneer in research on antihistamines, had published experiments showing that electrical stimulation of the brain leads to a breakdown of protein. Himwich and Samson were both intrigued by this report, and wondered if the extreme stimulation of diabetic shock would do the same thing. Given the previous experience of the Samson/Dahl team with diabetic research, Himwich thought it would be the perfect problem for them to tackle during the short time they would be in Galesburg.

Ungar's postulate had two components: When nervous tissue is stimulated mildly, the proteins undergo a reversible change in shape which could be detected, he claimed, by changes in the way that proteins absorb light at particular wavelengths. When nervous tissue was stimulated more vigorously, as surely would be the case during diabetic shock, the proteins would actually be broken down. Samson and the Dahls thus set out to determine whether a homogenate of brain tissue taken from cats induced by an overdose of insulin to go into diabetic shock would (a) show changes in its light absorbing properties, as claimed by Ungar, and (b) show a loss of protein, as would have to occur if protein were being broken down faster than it was being replaced.

Samson was excited. He was really getting at the chemistry of the brain and, by extension, the chemistry of behavior. He got even more excited when the first results were just as Ungar had reported: tissue extracts from the insulin-shocked cat brains showed changes in their light absorption properties compared to extracts from control brains. How fortunate he was, Himwich told him, to make such a dramatic discovery on his very first attempt!

To confirm their good fortune, Samson and the Dahls repeated their experiment. The results looked a little equivocal, so they repeated the experiment again. It began to appear that a lot of experiments were going to be needed to nail down the phenomenon with statistical certainty, so they settled into a routine of long, hard days in the sweltering heat of the hottest summer the three of them would ever recall, measuring the quantity and absorbance of batch after batch of brain tissue from insulin shocked cats. As the data mounted, their enthusiasm waned—the numbers seemed to be averaging out. The first flush of success they had seen in the remarkable differences between experimentals and controls faded into a background of inconsistent and unreliable results. The dramatic discovery that tantalized them at first seemed to melt like an ice cube in the August Illinois sun.

Norman Radin, a biochemist from the University of Michigan and a friend of Himwich, came to Galesburg for a visit. Would he be willing to talk to Samson about the strange results they were getting, Himwich asked? Radin agreed, and didn't have to see much to form an opinion.

"Why, this is nonsense! This is absolute nonsense! Where did you say you got this method from?"

"From a man named Ungar," Samson replied.

"From hunger?!" Radin retorted, making a pun of the answer.

A scientist of unquestioned technical competence, with an ascetic, humorless personality, Radin could be and was devastatingly negative and insulting. Samson had never seen a scientist of Ungar's apparent stature excoriated in quite that way before, nor had his own honest efforts met with such ridicule. But the vigor of Radin's critique made him step back from the work and start asking hard questions. Since the brain wasn't structurally decomposed or melted down by the functional trauma of insulin shock, the chemical changes induced by the experiment must be very subtle. And if they were as subtle as the partial unfolding of a few molecules, or the breakdown of a tiny fraction of the brain's protein pool, what were the chances that their crude analytical techniques could pick up such changes? Not very great, Samson began to believe. If a fish were taken from the ocean, would the water level change along the coastline? He began to wonder if the inconsistent results weren't reflecting random fluctuations of crude and imprecise measurements. Perhaps there weren't really any changes after all. Perhaps that first "lucky" experiment had just been a cruel statistical fluke. May-be Radin was right. Insensitive, but right.

Himwich saw it as a misunderstanding. Perhaps if Ungar and Samson could get together and talk over the experiments in person, they could resolve their inconsistent results. "Just talk it out," he suggested. "Maybe Dr. Ungar will have some ideas as to why your numbers aren't coming out." So at the meeting of the American Society of Physiologists that fall, Himwich arranged for Samson and Ungar to get together.

Samson's memory of the meeting was still vivid in reconstructing the conversation three decades later. "I started asking questions, and Ungar would just say, 'Well, I'm sorry; I guess you must be doing something wrong, because we always get this.' "

Samson pressed for details, his skepticism turning to anger. Ungar defended his own work with the confidence of the internationally known physiologist he had become, and puzzled over Samson's inability to repeat it. He was suave, cool, and in Samson's view, condescending. The more agitated Samson got, the calmer Ungar seemed to be. Samson brought up Radin, and the criticism of other biochemists. Ungar dismissed them. What did they know, if they hadn't done the experiments? But *he* had done the experiments hundreds of times himself, Samson argued, and he just couldn't confirm Ungar's results. The logical explanation, Ungar implied, was that Samson was doing something wrong. From Samson's point of view, their encounter was a total failure as far as any real exchange of information was concerned. He felt that Ungar essentially was saying, "Well, you're stupid. We're getting this result. I don't know why you are not getting it. Maybe you aren't doing the experiment right."

"We're doing it exactly as you described it in your paper!"

"Well, maybe you're overlooking something."

And so it went, never proceeding beyond the clash of a younger, unknown researcher presumed to be wrong, against the reputation and stature of an established scientist presumed to be right. Himwich deferred to Ungar's seniority; but he was a kind man, and could see that Samson felt humiliated and insulted. He tried to cushion the blow for Samson by being philosophical about the incident, and advised him not to take Ungar's attitude too seriously.

"You shouldn't get so emotional about this. You know—maybe you're right; maybe he's wrong. But he's older. He's more senior than you are. He's been in it longer than you." In other words, let the elder statesman have his way, but don't let the incident get you down. Don't take it personally.

How could he not take it personally, Samson thought? His integrity as a scientist had been questioned, his professional reputation impugned, and he should simply shrug it off? He couldn't. He wouldn't. And he never did.

Looking at the Samson-Ungar episode from the outside and at a distance, it isn't easy to understand the intensity of their dispute. When Himwich, Samson, and the Dahls finally published the results of their summer's work, it turned out that their experiment was not quite as precise a replication of Ungar's method as Samson had implied. The tissue homogenates that they made up for their measurements were two to five times more concentrated, and up to ten times less acidic than those prepared by Ungar. Also, they measured optical absorbance at a lower, less sensitive wavelength, and calculated the results differently. Above and beyond the methodological differences, however, was the fact that the experimental manipulations were fundamentally very different: Ungar had directly stimulated the nervous system electrically over a period of a few minutes, while Samson had fasted his animals, then given them hormone injections over a period of many hours. There was no inherent reason, in other words, to expect that the results should be very comparable.

In retrospect, it is tempting to speculate that the opinions of Radin and other biochemists had begun to weigh heavily on the physiologists (Samson and the Dahls) who were feeling their way into a new area without the confidence of experienced investigators. As Samson said, by the time of his meeting with Ungar, he had "become highly sensitized to [the opinions of] other people." Ungar, having no doubts about the validity of his own work, dismissed the implied criticisms of the biochemists and the physiologists in a way that Samson interpreted as condescending and insulting. In fact, what may have been more to the point was Samson's instinctive desire, as a young scientist, to come down on the right side of what he increasingly perceived to be a controversial issue early in his career. To be wrong is not a disaster, but to be associated with heretical ideas, as those of Ungar tended to become, can make the road to scientific acceptance much rougher later on.

This is not to detract from the basic validity of the results that Samson and the Dahls laboriously produced that summer. In coming to the conclusion that the analytical methods they (and Ungar) had used were too insensitive to pick up very small changes in protein content and structure, they were almost certainly correct. It would be another two decades before methods would be developed that were sensitive (and convincing) enough to demonstrate that the shape of proteins and even their breakdown can result from some types of excitation in the nervous system. In the long run, Ungar had the right idea, but in the summer of 1957, Samson's interpretation of the data was probably closer to reality.

Wiser and weary, Samson and the Dahls returned to Kansas, to the work they knew best—the work that had begun to make them known in the growing field of neurochemistry: energy metabolism in the brain. Their discovery of the striking inverse correlation between age and survival time in anoxia led them to investigate the brain's production of adenosine triphosphate (ATP), the primary chemical form in which energy is stored in most cells. They found that the brain of the rat requires much less ATP to keep it alive at birth than it does after it has matured. Since production of ATP is directly tied to the availability of oxygen, newborns are thus more tolerant to the lack of oxygen. Out of these studies came calculations of the cost in terms of energy that keeping the brain alive entails. Their research did much to illuminate the critical dependency of the brain on glucose (its fuel) and on oxidative metabolism (its means of converting the fuel to energy stored in the intermediate form of ATP.)

Dennis had decided to go to medical school. The first two years of basic science courses were taught on the Lawrence campus at that time, so he managed to keep in close touch with the lab. Nancy by then was a graduate student, continuing her research with Samson while struggling through the advanced course work of a doctoral program. In time, other students joined the team. Richard Lolley, who would later become well known in the field of retinal neurochemistry, joined the lab in the late 1950s, and did some of the first investigations on brain metabolism of guanosine triphosphate (GTP), uridine triphosphate (UTP), and other compounds chemically related to ATP.

Another addition at this time who would become a valued and semipermanent senior member of the team was William Balfour, son of a famous surgeon at the Mayo Clinic, and a Mayo grandson. Trained as a physician himself, he was struck after medical school by a long-term disability that required an extended hospitalization in Kansas. As part of his rehabilitation he had sought a low-pressure job in a research lab nearby, and Samson was willing and able to take him in. What began as a temporary rehabilitative measure, turned into a long-term position, then a faculty appointment, and eventually Dean of Student Life at the University. He was the perfect intellectual foil for Samson: calm where Samson was excitable, cautious where Samson was overly eager, organized where Samson had a tendency toward disarray. He became the trusted and competent lieutenant, and, as a practical matter, in many ways the operational head of the lab.

Through the early 60s, work in Samson's lab flourished. So many graduate students, undergraduates, and technicians were working in the cramped space available to Samson, that daytime and nighttime shifts had to be set up; and desks literally had to be stacked on platforms above one another to handle the personnel. The Samson and Dahl team flourished in the classroom as well. Still the showman, and abetted by Nancy's extroverted personality, Samson would do things like demonstrate to a lab full of medical students how to inject the tail vein of a rat. With Nancy holding the animal, he would stick the needle into the tail, push in the plunger of the syringe, then yank it out with a smile, while Nancy held the rat up to show it was no worse for wear. Most of the time, in fact, Samson's injection missed the tail vein, but the medical students couldn't tell it, and when they went to do it they were convinced from Samson's performance that it was easy, so they succeeded 9 times out of 10.

During site visits by external reviewers to evaluate his grant proposals on brain metabolism, Samson would impress his visitors by predicting precisely how many seconds a juvenile rat would take to regain consciousness following anoxic exposure to nitrogen. "Now that rat should roll over in about 70 seconds," he would say, as he continued to talk about other aspects of the experiment. Precisely 70 seconds later, the rat would open its eyes, roll right side up, and start scurrying about its cage, to the amazement of the visitors who hadn't heard another word that Samson had uttered.

The grants came in, year after year, bankrolling his expansive research effort. It was a good time to be in neurochemistry. Samson's emphasis on functional as opposed to purely descriptive chemistry of the nervous system kept his work near enough the cutting edge to merit continuing grant support. He also had a knack for shifting his focus after his prior approach had run its course. When he had mined about all he could from the fatty acid narcosis story, he shifted to the energetics of neural tissue and to the direct assay of ATP and other high-energy compounds. As the early 60s got underway, biochemistry was concentrating on enzymology, so Samson found himself studying adenosine triphosphatase, the enzyme that promotes the breakdown of ATP. The huge volume of information that we have today on the properties of that enzyme in neural tissue is due largely to the work of Samson, Balfour, the Dahls, and a long list of students, both graduate and undergraduate.

By 1964, Nancy and Dennis Dahl were out of graduate school and medical school respectively. Meanwhile, Samson had been promoted; and when the medical school moved from Lawrence to Kansas City, he became chair of the remnants of the biochemistry and physiology departments that stayed behind.

Neither the biochemists nor physiologists wanted to merge with the zoologists (mainly taxonomists) at the Museum, so the administration contrived a department that only an administration could love: a Department of Comparative Biochemistry and Physiology. Samson could have moved to Kansas City with the medical school, but he liked the ambiance of Lawrence and elected to stay. His old friend and chairman, Ken Jochim, wanted to go to Kansas City as chairman, but wasn't asked to; and refused in a huff to stay on as chairman in Lawrence. Thereupon, Samson was asked to be interim chairman of the newly created department in Lawrence, and in time that appointment became permanent.

For Nancy and Dennis, the time to spread their wings was long overdue. Sensing that, they applied for fellowships to go to England to work with Derek Richter, another neurochemist very interested in functional problems. But their year in England was not marked by any notable scientific successes, and was overshadowed by a deep personal loss. After 12 years of marriage, Nancy became pregnant, but the male child that was born to them survived only a couple of days. Depressed by that, and homesick, the Dahls made the fateful decision to return to Kansas when their year was up, rather than reapply for fellowships to keep them in England. Samson and Balfour had written that funds were available to support them back in Lawrence; and as always, Samson had a host of ideas that needed some enthusiastic hands—by then, especially, since administrative duties had effectively taken him out of the lab. In retrospect they would later conclude that it was a mistake to go back to the comfort of the only scientific milieu they had ever known. They were both at a point in their careers where they could have ventured out and probably succeeded on their own. But at the time, it seemed the right thing to do.

At some point during their stay in England, they came across a small publication called *The Neurosciences Research Program Bulletin*, published by a Boston organization associated with the Massachusetts Institute of Technology. It was not a scientific journal in the usual sense, but more like a newsletter of an organization vaguely resembling a think tank devoted to the operation of the nervous system. They had no idea what the organization was really about, and no knowledge of its founder, Francis O. Schmitt. But they knew Samson well enough to know that he might be interested, so they brought the *Bulletin* to his attention, once they were back in Kansas. As it turned out, that act was the last great impact the Dahls would have on Samson's career. Ironically, the events that flowed from Samson's acquaintance with the Neurosciences Research Program contributed substantially to the wedge that grew between him and them in the years that followed.

As the momentum and reputation of Samson's research continued to build, he got an NSF grant for Undergraduate Research Participation, which ideally suited his propensity for engaging students in research. Samson drew from science departments from all across the University, making it one of the biggest Undergraduate Research Participation programs in the nation. He had a unique philosophy about how the funds should be awarded, though—a philosophy that

NSF severely frowned upon when it found out about it later. Samson noticed that the projects selected for funding by NSF varied radically in quality from year to year, with no reason or consistency. Luck seemed to play a major role in whether the reviewers in any given year preferred the projects proposed by the physics department, say, or the zoology department, as opposed to those from other departments which happened to draw unfavorable reviews in that particular cycle. Samson decided, therefore, to share the wealth—to disregard NSF's selection and to administer the funds as though awarded to the University as a whole. That way, the luck of the draw for one department and the misfortune of another were smoothed over from year to year, and a consistent level of support was provided for all.

NSF understandably felt that Samson had undermined its peer review process, and removed him as the project coordinator at the University of Kansas eventually. But before they found out about his Robin Hood style of administration, they gave him glowing reviews. In the early '60s, NSF pointed to the Undergraduate Research Participation program at Kansas as a model for the nation.

Samson made another controversial administrative decision, though this was less critical and it involved his own Dean rather than NSF. From the start, Samson thought it would be good to bring students from outside the University to the Lawrence campus for the summer programs. The Administration was more inclined to restrict awards to students already enrolled. NSF agreed with Samson on this one, so from time to time, students from other colleges would come to Lawrence to take part in Samson's vigorous and expansive program of undergraduate research. That is how I found myself in Kansas in the summer of 1964.

Living the Dream

I arrived in Lawrence on the rainy afternoon of June 2, with one heavy suitcase, little money, and no place to stay. I found a room for $7 a week at the base of the hill below the University, where I spent most of the next three days reading about energy metabolism and the brain. On the fourth day, I went to meet the author of the work himself.

Frederick E. Samson was not quite what I had expected—a little shorter in stature, decidedly less professorial looking, and a lot more talkative. It was a Saturday morning and he was working in the office by himself, in a short-sleeve sport shirt open at the collar, chewing on a cigar. Since I had approached this first encounter with some anxiety, his ability to put me so readily at ease made a great impact on me. He showed me through the labs, assigned me to a desk of my own, and talked with great animation about the work that he was doing. To my delight, I was well enough prepared to follow what he was talking about, and to engage in a real give-and-take conversation. By the end of our first session together, I suspect he was feeling some relief at the thought that the chance he had decided to take on the outside student from Texas Tech was probably not going to come back to haunt him.

By Monday morning I was ready to go to work, but the electrical engineering student with whom Samson had assigned me to work was still on vacation, so I did more reading, which led to more hypothesizing. It occurred to me that memory might involve somehow the relative enhancement or inhibition of enzyme activities related to the energy metabolism that Samson was studying. Already I had adopted the tendency that later would come to characterize the field of research on biochemical correlates of memory in general: ascribing a role to whatever happened to be the molecule of the moment in the investigator's mind to some critical mechanism of memory storage.

Tuesday morning, Samson rescued me from my speculative reverie by showing me a recent scientific paper that claimed to reverse the anesthetizing action of procaine through the application of high-energy compounds such as

ATP. For some time it had been known that procaine (Novocain) was able to block pain in a local area by blocking the ability of the nerve to depolarize, or to conduct a bioelectrical signal from one point to another. This paper claimed that a molecule like ATP could undo the block, probably by some type of chemical reaction with the cell membrane. At that time, it was still relatively unusual to think of nerve excitation in terms of molecular, as opposed to electrical, mechanisms. It was a phenomenon tailor made for Samson's interests, and, by extension, mine.

When Frank Scammon, my research partner, arrived on Wednesday, we went to work immediately on trying to replicate the previously reported findings, using as our tissue for experimentation the vagus nerve of the turtle. Since Nancy Dahl had pioneered the use of the vagus nerve from the chicken for this type of research, the biological methodology was already worked out. Frank knew all the electrical instrumentation, and I knew enough biochemistry and physiology to handle the solutions and tissues, so we were running actual experiments by the end of the week. It was hard work, but I was so excited I could hardly contain myself. Away from the West Texas plains for less that two weeks, and I was already doing big time research in neurophysiology! On Friday, after a week of work in Kansas, Samson reviewed what we had done, and said it was good.

The days rolled on. Every morning I awoke with a sense of mission and adventure. I couldn't wait to get to the lab. And I was good at what we were doing. I couldn't have managed the electrical instrumentation on my own, but I learned from Frank quickly and well. This was so different from the painstaking field work in zoology, where it took at least a summer to amass some data that might or might not fit into some coherent pattern. By contrast, a single day's work in neurophysiology conceivably could result in a clear answer to a precise question. (And, not incidentally, it could be done in air-conditioned comfort.) So every day I awoke to the possibility that this might be the day of the breakthrough—the big discovery that would tell us something new and fundamentally important about the function of nerve cells.

Frank seemed to enjoy our work, but he wasn't mystical about it like I was. It was a job for him, and when it was quitting time, it was time to quit. My quitting time was whenever the experiment naturally ended, and that could be well past the dinner hour. Frank upset me greatly on a couple of occasions when he stopped in the middle of an experiment simply because it was 4 o'clock. I worked right through lunch as often as not, and would work all night willingly if that was what it took.

In the end, the summer fell short of my expectations in a couple of important respects. Despite the optimism of our exciting beginning, the more data we collected as our research project progressed, the more unclear the results became. Up to the last week of the summer, I held out hopes that we would learn enough about the interaction between nucleotides like ATP and local anesthetics like procaine to publish a paper on the chemical aspects of nerve excitation, but the clarity didn't come.

On the other hand, I had confirmed to myself that I could do physiological research and that I really enjoyed it. Secondly, I had come to see in Samson a model of what a scientist should be like. He appreciated long hours and hard work, but effort alone was not its own virtue. There had to be a good idea behind the effort. Imagination counted with him. Maybe imagination counted to a fault, but it was a welcome correction to the relentless glorification of toil and sweat that I had experienced as an undergraduate working in field biology.

At the end of the summer, I took the train to Chicago, to visit friends and check out the University of Chicago—Samson's alma mater. Chicago was the biggest city I had ever seen, and I was awed to speechlessness by the taxi ride along Lakeshore Drive. The palisade of sky-scrapers, the density and bustle of traffic, the noise and odor and energy of an urban sprawl more vast than any I had ever imagined, contrasted so starkly with the lazy college town of Lawrence and the diminutive city of Lubbock, that my senses were overwhelmed. On the other hand, I found the close-spaced gothic buildings of the University of Chicago campus, and the vastness of the ghetto surrounding it, both rather intimidating. But the big city on balance won me over. If I ever had a doubt about needing to leave Texas, the trip to Chicago left me knowing that a larger world was out there, waiting for my arrival.

First, there was the matter of my senior year at Texas Tech. And with graduate school looming on the horizon, the time to apply had come. From a list of over twenty possibilities, with excitement and guarded optimism, I paired my choices down to four by October: Western Reserve, Yale, the University of Pennsylvania, and the University of Kansas, which was, of course, a safety school. I had enjoyed my experience there tremendously, and had no doubt that the graduate education there would be fine. But I really wanted to go somewhere new, to work with scientists whose research was more behaviorally oriented and closer to the exciting developments that I knew were probably just around the corner. My grades were not great, however, and the graduate schools across the country were flooded with applicants, so a touch of common sense told me to apply to a place where I was known as a real person instead of a transcript and a GRE score. Not only did that turn out to be a prudent decision, since I was rejected by Yale, Penn, and Western Reserve (in that order), but the day after my rejection from the latter, my worst fears were realized when I was informed that no assistantships were available for any first-year student in my home department at Kansas. From the high aspirations and hard work of the fall and winter application process, I had managed to get into only one graduate school—my fourth choice of four—and with no financial aid even there.

The Gospel According to Schmitt

By 1964 Francis Otto Schmitt was a man who had lived long enough to generate several reputations and a couple of disparate images. A gee-whiz enthusiasm for science and a can-do boldness in both his methodological and theoretical approaches would characterize his entire career, befitting his Midwestern American origins. He was born and raised in St. Louis, where he received his bachelor's degree in 1924 and Ph.D. in Physiology in 1927 from Washington University. He grew up in science during an age of giants, when great strides and fundamental discoveries were made by a relatively small number of notable individuals. This experience, plus postdoctoral studies in Europe, engendered a Teutonic attitude that didn't wear so well in later years when scientists became a dime a dozen and breakthroughs came too often from unknown quarters.

To those who knew him as a young professor rising quickly through the academic ranks at his alma mater during the 1930s, he was an innovator at the cutting edge of technology and a scientist of the first rank. Those who came to know him at the pinnacle of his research career as Chairman of Biology at the Massachusetts Institute of Technology during the '40s and '50s saw him as one of the nation's outstanding researcher/administrators of the post-World War II era. Not only did his work on the ultrastructure of biological tissues, based largely on pioneering studies with the electron microscope, pave the way for visualizing biological structures as molecular entities, his personal ties with men of status and power throughout the governmental and academic establishment enabled him to influence the direction of entire fields of study. By 1964, he was largely out of research and into his role as full-time impresario of the scientific (and to him this meant mainly the biophysical) study of brain function. He had founded the Neurosciences Research Program (NRP) in 1962 as basically a highly elite think tank aimed at proselytizing the gospel of the coming glories of research on the biophysics of the mind. In this more mature phase of his career, his image struck different people in different ways. At a visit to the University of Kansas in 1964, he left Fred Samson in awe, and impressed Nancy and Dennis Dahl as a big bag of hot air.

The dust was still settling that year from the move of the Physiology Department to the medical school in Kansas City and from Jochim's refusal to remain as chairman of the residual department in Lawrence. Samson had been drafted to fill the breach, and his grateful Dean gave him a thousand dollars to spend as he pleased as a token of appreciation for his handling the debacle as well as he had. The Dahls had brought back word of the existence of the NRP from England, where they had come across an early mimeographed copy of the organization's house organ, the Neurosciences Research Program Bulletin. The NRP concept excited Samson, and, remembering back to his graduate school days when Schmitt had received an honorary degree from the University of Chicago, he decided to use the Dean's money to bring Schmitt and a couple of other noted scientists to Lawrence.

The substance of Schmitt's talk and visit in Lawrence is not remembered clearly by anyone, but the impression he left was indelible and totally contradictory for Samson and the Dahls. Schmitt's evangelical zeal for a new, over-arching, integrative approach to the study of the brain hit Samson's restless but fertile and prepared mind at just the time when he was beginning to look for a new creative twist to his career. The Dahls, on the other hand, being more firmly grounded in the practical realities of day-to-day experimentation, saw Schmitt's message as vacuous and his manner as pompous. In reminiscing about the experience years later, both Dennis and Nancy were unsparing of both Schmitt and Samson in their recollections.

"F.O. came here and I was totally unimpressed," recalled Dennis, referring to Schmitt by his first and middle initials in the manner of all who found the need to distinguish between Schmitt (F.O.S.) and Samson (F.E.S.) in later years.

"He came here and gave a seminar, then we went over to the Union for dinner, and F.O. sat there and dropped names, and dropped names, and dropped names . . .'Ah, yes, when I was meeting with blah-blah this last week. . .' "

"And often it wasn't just 'blah'," Nancy added; "It was 'Nobel laureate blah-blah!' "

"We sized him up with our first meeting," continued Dennis, "but Fred was just overwhelmingly impressed."

"Fred sat there and hung on every word," according to Nancy. "I couldn't believe it.

Dennis summarized on a note of reluctance.

"I just couldn't believe that he (F.E.S.) would be so taken in—that he would be so gullible, just because the guy (F.O.S.) dropped a bunch of names. . . . All of us have idols, and eventually we find that our idols have clay feet. Fred to both of us was an idol. Fred to both of us was somebody who was really fantastic. I still think he's fantastic. But I think when he first met F.O. was when I found he had clay feet."

The harshness of the Dahls' recollection makes more sense in the light of subsequent events, which saw both Dennis and Nancy passed over for faculty positions or promotions that they think Samson should have fought for on their behalf, and in the salary cuts necessitated by grant funds that dwindled as Samson was drawn by Schmitt into the NRP orbit, both literally and figuratively. Not that their judgment hasn't been shared by others familiar with Schmitt's flight-of-fancy rhetoric coupled with an incessant reflexive reference to authority. But one could also argue that Schmitt's impact on neuroscience justified much of his rhetoric, and that the names he dropped in pursuing his objectives are names which indeed accounted for the success of his later years as an impresario of first biophysics, then neuroscience. In faulting Samson, too, the Dahls may well have misjudged the basis for his enthusiasm. Given his predilection for grand schemes, his more theoretical biophysical training, and the fact that his outlook was primed for a transition, it seems more likely that it was the substance of the message rather than the messenger or his manner that struck such a positive chord in Samson's thinking.

Whatever the reason, Samson decided that he had to go to the NRP. With the University having owed him a sabbatical for some time, he asked Schmitt if he could come to Boston as a Visiting Scientist for a year. Schmitt agreed and Samson got a stipend for the purpose from NIH. By September of 1965, he had been installed in a tiny back office at the NRP's home above the American Academy of Arts and Sciences on the Brandegee Estate in Brookline, Massachusetts. There he floundered, alone and ignored by Schmitt for six or seven of the most agonizing months of his career. With no laboratory experimentation to focus on, and with no duties or assignments to occupy his time, all he did was watch the parade of world-class researchers who passed through the building, impressing Schmitt and one another with their scientific and intellectual virtuosity. Both in quality and in quantity, the science that was bandied about the meeting rooms of the NRP was intimidating to Samson in the extreme. Even if he had been working near the cutting edge of neuroscience, it seems unlikely that he could have remained unfazed by the number of scientific luminaries that came to call at the house that Schmitt had built. He felt overwhelmed and became depressed. He wanted to go back to Kansas, but decided to stick it out for a year.

One day Holger Hydén, the Swedish neurochemist then considered one of the foremost researchers in the learning and memory field, spoke on a university campus somewhere in the Boston area, and by chance Samson ran into Schmitt at the talk, so they sat together. At the end of the talk, Schmitt asked Samson if he wanted to go have a beer. Samson was floored, and of course accepted. F.O.S. and F.E.S. had their first real conversation together since Samson had come to Boston. In the days that followed, Samson found himself in Schmitt's office with increasing regularity. A chemistry between the two had begun to brew.

The first big project that Samson played a key role in was a work session on axoplasmic transport, the phenomenon by which chemical substances produced in the body of nerve cells are carried down the long extensions of the neuron to its terminals. Just as blood is pumped from the heart to the extremities of the body for the delivery of oxygen and nutrients, the distant terminals of nerve cells depend on a transport system for their supply of nutrients and functional molecules manufactured in the central cell body. It was thought by the early Sixties that this transport depends on long filamentous proteins called microtubules. Axoplasmic transport had first been studied in detail in the earlier part of the century by Paul Weiss, a towering figure in the field of developmental neurobiology. He would have been the natural pick to organize a work session on the subject, except that his ego matched his stature and was exceeded only by his contentiousness. Schmitt knew it would be a disaster to try to have Weiss either organize or chair the meeting. In a purely political move, therefore, he tapped a young medical doctor turned researcher at the McLean Hospital who had just begun to study the axoplasmic transport process in the brain, rather than in the peripheral nerves as Paul Weiss had done. For Samuel Barondes, it was the second great break in his career—giving him great visibility in a hot, emerging field on top of the publicity that his recent work on learning and memory had garnered.

For Samson, the Work Session came at a critical time as well. He had never done any work on microtubular protein, but his exposure to the problem at NRP gave him an inside track that he would later exploit when he returned to Kansas. Of equal or even greater importance was the style of interaction between him and Schmitt that evolved in the course of preparing for the Work Session. Though Barondes was the titular organizer and chairman, Schmitt showed the still youthful psychiatrist how to do it; with Samson serving as the unseen staff lieutenant behind Schmitt's forceful, guiding hand. F.O.S. and F.E.S. developed a method. Schmitt would get on the phone with someone, and Samson would be on the extension, not speaking himself but passing notes to Schmitt during the conversation, then analyzing with Schmitt the meaning and effect of the conversation after it was over. F.O.S. would bombard F.E.S. with half-baked ideas, parts of talks, quasi-correct factual data, and chaotic ramblings; then F.E.S. would throw it all back, rebaked, semi-critiqued, quasi-corrected, and sometimes more, sometimes less coherent. The cycle would repeat itself. Like a whirlpool in the bathtub that spirals toward finality, the facts and ideas would eventually resolve into an action, or talk, or paper that was reasonably clear, coherent, and relevant. But to the outside observer watching F.O.S. and F.E.S. in action, no process appeared more mysterious or incomprehensible. It was a measure of the power of their particular interaction that no one else could understand it, nor did anyone else ever emulate the role that Samson played with Schmitt to a very successful degree.

The Work Session on axoplasmic transport was a great success. It established the credibility of some important new findings, like transport within the central nervous system and transport at a much faster rate than Weiss had believed; it introduced a host of new scientists in the area; and it propelled the study of axoplasmic transport, microtubules, and their proteins into the forefront of research in neuroscience. For Barondes, it once again thrust him into the limelight of a new field at just the right time. For Samson, it stimulated a change in his research that he would later come to regard as his most enjoyable and rewarding work. When the NRP Bulletin on Axoplasmic Transport came out, it was authored officially by Barondes, with an essay by Schmitt, but Samson's effort pervaded the work, and Barondes himself acknowledged the importance of Samson's contribution.

By the end of Samson's year at NRP, Schmitt had come to rely on him so much that he wanted him to stay. With the indulgence of the University of Kansas, Samson agreed to another year with Schmitt. Nancy and Dennis Dahl were not pleased. And a graduate student who had come to the University of Kansas to work with Samson was gravely disappointed. What I didn't know at the time was that I was one of the reasons, if only a small one, that Samson had become valuable to NRP. And that had to do with events that unfolded in Houston in the summer of 1966.

Ungar

"You Will Never Amount to Anything in Science"

Georges Ungar was born the only child of a French Catholic mother and a Hungarian Protestant father in Budapest on March 30, 1906. He was raised in France, where his home environment was intellectually stimulating, but not oriented toward science; so his early interests and talents led him to excel in the humanities. He won first prize in Greek scholarship in France in 1925, and finished second in History to a person who later would become a noted historian. At the threshold of his university studies, he really wanted to be an art historian or literary critic, but recognizing the difficulty of making a living in those professions, he decided to follow his father's lead into architecture.

The problem with architecture, he discovered to his dismay, was that it relied on more math than he knew or cared to learn. So for reasons he could not recall in later years, he decided to become a physician. His initial training was not auspicious; the basic science curriculum required of medical students at that time in France did not interest him, and the cultural and (one assumes) social diversions for a young bachelor in Paris were considerable. His academic performance was poor, and inspired no predictions of success. His zoology professor, upon learning of his decision to go into medicine, told him "It is just as well, because you will never amount to anything in science; your drawings are terrible."

His interest picked up with the beginning of his clinical studies in the second year. The hospital wards in the 1920s were not pleasant places. Few truly effective drugs were available, and medical science was still at an early stage of development. He was appalled to learn how little could be done for patients with serious diseases. For the rest of his life he was particularly haunted by the poignant case of a beautiful young girl who died slowly of bacterial endocarditis. The interns all fell in love with their patient, and followed the course of her disease with a neurotic fascination, as the pallor caused by her condition enhanced her attractiveness. When she died, Ungar was among those required to perform her autopsy.

In spite, or perhaps in part because of, experiences such as these, he became inured to the suffering around him enough to develop the objectivity necessary for making accurate diagnoses. He relished the puzzle-solving methodology of this aspect of medicine, and began to distinguish himself as a diagnostician. However, his bedside manner left something to be desired, so it became apparent early in his clinical studies that, contrary to his zoology professor's prediction, he would amount to more as a scientist than as a physician. The science of physiology, then as largely still today, was considered the most basic science of the medical profession. The physiologist before the Second World War was regarded with a bit of the glamour associated with the molecular biologist of today. It was to this enviable profession that Ungar began to aspire. Through a friend of a friend of an eminent professor's relative, he was introduced to the eminent professor, and found satisfaction as a physiologist for the rest of his career.

Physiology is the study of the function and control of the component parts of an organism. The anatomist studies how the body is put together. The physiologist studies how the parts of the body carry out their functions and control one another. Physiology grew out of the mechanistic views of the Enlightenment. France was fertile ground for the Enlightenment, and produced two of the greatest scientists in the history of experimental medicine. Everyone knows of Louis Pasteur, and the importance of his empirical, experimental investigations that proved the microbial origin of disease, disproved the spontaneous generation of life, and founded the basic tenets of microbiology. Less well known but of equivalent importance to physiology was Claude Bernard, who made a number of important discoveries relating to neural regulation of blood flow and mechanisms of digestion, but more importantly showed the necessity of careful experimentation on living animals for under-standing how the body works. Befitting its origins in the mechanistic thinking of the Enlightenment, Bernard's approach was extremely mechanistic and empirical. His contemporaries, if not he himself, stretched this thinking to the limit, as illustrated by a story told about a neurophysiologist, Francois-Franck, who studied with Bernard. One day Francois-Franck showed a visitor into his laboratory, where a disemboweled animal had tubing into several veins, its heart attached to a cardiograph, and one of its kidneys enclosed in another instrument. "Now we are ready to study emotion!" announced the physiologist.

Francois-Franck had a student, Jules Tinel, who became an accomplished neurologist, and it was Tinel who offered Ungar his first independent laboratory position. This was the link that would enable Ungar to claim intellectual descent from Bernard himself; as did most of the world's physiologists in time. Like a majority of the research labs of that period, the equipment was meager and the facilities rudimentary. Furthermore, Ungar received virtually no salary, but it seems that he didn't need much. He was happy doing the methodical work of the researcher, testing his notions of how animals function. He would later speak of the pleasure that scientists invariably point to, independent of any broader benefit (or detriment) to society. There is great personal satisfaction that comes from watching an idea be confirmed or rejected, as the points accumulate on a graph, or the numbers build up in a table. Even if, as often, the results are equivocal, you know you are looking at information acquired by your own ingenuity; enticed from nature by your own skill and design.

Ungar's choice of Tinel as a mentor anticipated two fateful characteristics that both men shared. Tinel was not much of a political manipulator, and Ungar had no taste for this aspect of academic life either. That both men lacked this skill probably worked to the detriment of their careers. Secondly, Tinel was attracted to some peculiarly complex problems at the interface of neurology and psychiatry—hysteria, neurosis, and psychosomatic disorders, for example. Eventually, Ungar too would pursue difficult problems that straddled conventional disciplines in an uncomfortable way. He would study the chemistry of excitation when most physiologists thought of excitation as an electrical phenomenon. He would view drug addiction insistently from a physiological as well as a pharmacological perspective. And eventually he would search for memory as an interplay between the highly ordered circuitry of the brain and its disordered molecular composition. To take an interdisciplinary approach to any problem is respectable enough—indeed is lauded by neuroscientists whose subject matter today has come to require it. But it invites criticism from two or more fields, rather than just one. If the research has high visibility and important implications for others working in peripheral fields, the criticism, both legitimate and illegitimate, has a greater chance of catching hold.

Whatever the long term effects of his association with Tinel were to become, the short-term effect for Ungar was recognition and growing fame. Tinel believed that histamine might be involved in the dilation of blood vessels that was often seen in nerve-controlled diseases and certain psychosomatic conditions. The idea was not original with Tinel, but he tried to prove it experimentally. Thus Ungar was put to work trying to define substances released from nerves by stimulating them in a variety of ways. By the late 1920s, it was known that histamine is a naturally occurring substance in many cells, and that its release is elicited by cell injury or antigen-antibody reactions. Histamine causes the blood vessels to dilate (hence the "blush" of the skin as more blood flows near its surface), causes itching by stimulating sensory nerves, and promotes contraction of many smooth muscles, such as those of the intestinal wall and the bronchial tubes into the lungs. That is why antihistamines are taken to open the airways (by relaxing the smooth muscles surrounding them) during an asthmatic attack. Ungar developed the notion that the variety of defensive reactions of the body to any injurious stimulus were all mediated by the release of histamine from cells of the nervous system, and appears to have been the first to use the term "histaminergic" to refer to the class of nerve cells for which histamine is a neurotransmitter. This turned out to be incorrect, but in the process of testing the idea Ungar developed the reputation as a leader in the field of research on histamine.

Central to Ungar's work at the time, and a key to understanding his research approach to the end of his career, was the procedure of the bioassay. A bioassay is a measure of the amount of a substance based on a quantifiable biological response to a mixture containing the substance. If, for example, you want to know how much of a substance is present in a certain tissue that will cause a muscle to contract, you dissect out a muscle that you know will respond to the substance, hook it up to an instrument that measures the extent of contraction, and apply the mixture to the muscle. By recording how much the muscle contracts, a quantitative inference can be made about the amount of the substance causing the contraction.

Bioassays have been tremendously useful in the early stages of research on naturally occurring substances, before they have been chemically defined, when the extent of their presence or absence in specified tissues under a variety of circumstances is the focus of study. Almost every hormone we know about was discovered and studied at early stages by this method. It was discovered early on that the smooth muscle of a segment (the ileum) of the small intestine of the guinea pig is very sensitive to histamine, so the degree to which the guinea pig ileum contracts was adopted as a standard bioassay for histamine. Ungar used this technique in testing for the presence of histamine under the different conditions that he suspected caused its release. His strategy depended on making sure that the muscle was responding only to histamine. In order to prove this, he began testing experimental drugs that might block the contractile effects of histamine. Daniel Bovet from the Pasteur Institute gave Ungar a variety of synthetic chemicals, one of which turned out to be weakly but specifically antagonistic to histamine, and to contain the chemical structure that later would be shown to be necessary for blocking the action of histamine. Bovet, who went on to discover other, more potent antihistamines, and to win the Nobel prize in 1957, was generous in acknowledging Ungar's early work on this class of drugs that brings relief from itching, inflammation, allergies, and cold symptoms to all of us from time to time.

Though the antihistamines were a great success for medicine and the pharmaceutical industry, they were disappointing to Ungar because they showed that histamine was not the only substance released by injurious stimuli or allergic responses. In the process of disproving his own hypothesis, though, Ungar established an international reputation, and his lab became the place to go to learn to do histamine research. It was for this reason that Alberte Levillain came to him, to learn the bioassay for histamine, in March of 1937.

Alberte was born on July 4, 1913 in Indochina, the oldest of three daughters to a schoolteacher and a former French naval officer who became harbormaster of the commercial port at Saigon. Her father reenlisted when war broke out in Europe a little over a year later, so the family spent the war years in France but returned to Saigon in 1920. In time the Levillains decided that their daughters should return to France for their education, so Alberte was an accomplished traveler by the time she was a teen-ager. She read from as early an age as she can remember, and was drilled in German and the classics to high standards by her grandfather. For a time the family lived in Marseilles, which the girls found oppressively dreary after the tropical ambiance of Saigon; but by the time Alberte was ready for the university, her mother had decided to settle in Paris. Alberte found this city distinctly more to her liking. She wanted to study medicine, but was discouraged from doing so because of her gender. Pharmacy, on the other hand, was considered entirely appropriate for a young woman, so long as she was prepared to get married at the first opportunity. The educational program was about as broad and rigorous in the basic sciences as was that for medicine, so she took that path with no apparent regret, acquiring her license to practice pharmacy by the age of 23.

She was interested in pursuing research, however, and managed to get an appointment through the Pasteur Institute with a bacteriologist who was interested in finding out if microorganisms produce histamine. The bacteriologist was a senior researcher much impressed by a young physiologist from Tinel's group who had mastered the technique of the bioassay for histamine. Thus Alberte Levillain was sent to Georges Ungar, to learn how to determine if bacteria produce the substance that causes the guinea pig ileum to contract.

At the age of 30, Georges Ungar was already bald and not the most handsome man, but attractive, in Alberte's view. She was first surprised that a man of his reputed accomplishments would be so young. And she was impressed by his clear gray eyes that were really "quite beautiful." He was pleasant to her, displaying a congenial personality and his considerable intellect from the start. But the truth was that she was not interested in him socially in the beginning, because she was nursing at that time a broken heart from a serious romance that had not worked out. She was not ready to get involved again so soon.

The same was not true for Ungar. It can be inferred that his bachelor years in Paris to that point had not been socially wasted. His sex life had begun with a woman ten years older, a friend of his mother, when he was 17. He had serious long-term relationships with a couple of other women, including a recent one just prior to meeting Alberte. By the time he met her, he was primed for marriage because he began talking about it "practically within a week." They began their collaboration; she worked in the hospital pharmacy every morning and on research with him in the afternoons. He took her out, showered her with wit and personality, impressed her with his knowledge of science and just about everything else, and began to win her over. He was up front about what he wanted.

"I have to marry either a rich wife, or a wife who works." he told her.

"That's all right; I'm not rich but I'm willing to work," she replied.

His other condition was not so easy: "I don't particularly want children." She was a bit shocked, having been raised to consider children an inevitable correlate of marriage. But she clearly wanted to be more than a housewife and had professional ambitions of her own, so it didn't matter to her so much at the time. With the ground rules thus established, and with war impending, they got married in October, under seven months after they first met.

Ungar did not enjoy frivolous vacations. Before he was married, he had learned to relax, have fun, and get some work done at the same time by spending his summers at marine biological stations, including the one at Arcachon on the Atlantic coast of France where decades later Stanislav Tauc and Eric Kandel would perform critical experiments on cellular mechanisms of plasticity in marine mollusks. Most scientists enjoy this casual mix of work and recreation. Thus an inordinate number of biologists leave their home labs and classrooms for extended periods for projects by the sea. The first few summers that Georges and Alberte were together, they spent at marine biology labs on the coast of France. She could not remember what they studied in any detail, apparently recalling the pleasure more vividly than the practical purpose of their retreat. These interludes came to an end, of course, as war descended in 1939.

Ungar joined the French army, only to find that a doctor of science rated no special treatment. Years earlier he had finished his medical studies but become so deeply involved in research and sure he would never practice in the clinic, that he had simply not bothered to take his final licensing exams. The prospect of facing Hitler as a private gave him sudden incentive to take his finals, "which of course he passed brilliantly," according to Alberte. As a medical doctor, he quickly was promoted to officer, and by the time the "phony" war began, was working in a chemical warfare and munitions center south of France. Alberte remained in Paris, promoted to chief of the pharmacy as her male superiors were mobilized; but she could come down to his lab to visit on weekends. The war turned real in the course of such a visit, and getting back to Paris as the Germans approached was not feasible, so they headed south together. The armistice came swiftly and Ungar was discharged from the army but, like all French citizens, forbidden to leave the country.

Though they could have returned to their former homes in Paris, they chose to stay in the unoccupied zone and search for a way to heed Churchill's call for patriots of France to come to England and join the allied cause. Through Algeria and Morocco they managed to reach Gibraltar illegally in March 1941. English scientists who knew Ungar's work helped get visas for him and Alberte. The final leg of their escape was the hazardous voyage by convoy from Gibraltar to England, escorted by warships, under occasional attack from German submarines. In London, they were designated as official members of His Majesty's Allied Forces. As French citizens they also enrolled in the Free French Forces of DeGaulle. With his presumed knowledge of chemical warfare and her pharmacy experience, they were assigned the task of distributing gas masks—an apparent compromise between their scientific expertise and lack of facility in English.

After a few months of this daytime boredom and the nighttime anxiety of frequent air raids in London, Ungar was asked by Solly Zuckerman, the distinguished biologist, to join his research effort at Oxford. Here the Ungars enjoyed a quiet and pleasant life, as no bombs fell on Oxford, and the intellectual stimulation of that university community was much to their liking. Ungar found himself the pawn in a feud between Zuckerman and LesGros Clark, an eminent physical anthropologist, for a time, but otherwise the social climate was congenial as well. Within three months, the Ungars became conversant in English. This Alberte realized one day when she found with great satisfaction that she could understand the gossip from the seat behind her on the bus.

Ungar's research at Oxford was concerned with traumatic shock and inflammation. The work consisted of exploding structures housing animals, then studying the effects of such trauma on the animals that survived. Surely it was research of a gruesome sort, but the values of peacetime don't translate well to periods when humans are blowing themselves apart. One of the dubious but unquestioned benefits of war is that they have accelerated our understanding of how to deal with trauma and treat disability. In the broader context of basic science, this research extended Ungar's work on defense mechanisms of the body, especially with regard to inflammation. His results anticipated the anti-inflammatory properties of the adrenal corticosteroids, such as cortisol—commonly used today for the control of itching and inflammation. The definitive research in this area was being done in Canada and the United States, unknown to him at the time. He also began work on the novel idea that the spleen might also be involved in the body's defensive mechanisms.

When the war ended in 1945, Ungar initially expected to go back to Paris, but stayed in London to serve at the French embassy as a science advisor. In this capacity, he was able to arrange a trip to America, where by then he had heard that considerable advances were being made in understanding the role of the pituitary and adrenal hormones in stress and inflammation. Alberte was coming to grips with a more personal issue. If they were not going to have children ever, then fine, she told Ungar; but if they were, they should "get on with it." By this time he had mellowed from his cavalier pronouncement at the start of their marriage that "A woman who has had a child is no longer a woman." Now seven years later he had come to the opinion that it would not be fair to tell a woman who wanted a child that she could not have one; so they decided to try, but "just for one." They were not long in waiting. Catherine Ungar, their only child, was born in London on May 12, 1946.

Crebiozen, Phenformin, and Resurrection in Houston

At the age of 41, Georges Ungar embarked on a new phase of his life and career in the New World. Years later he would still remember his arrival in the United States with obvious pleasure.

"The famous skyline of New York harbor surpassed my expectations. Many things were new and exciting to me: the awesomeness of the tall buildings, the splendid vulgarity of Times Square, the colors of the fall in America and, later, the violence of a snow storm paralyzing a big city."

His research was well enough known on this side of the Atlantic to merit invitations for him to speak at numerous institutions, including Yale, Cornell, and Johns Hopkins. In the course of these visits, abetted by interest in his work on inflammation, he was offered several positions. While he had not originally intended to stay in the United States, he must have sensed the opportunities that this nation would bestow on the scientists of the postwar era. So he accepted the most attractive of his job offers: a position as research scientist at the Northwestern Rheumatic Institute, located in the Cook County Hospital in downtown Chicago.

He picked up where he had left off in London with research on responses to traumatic injuries, especially burns, and on the possible involvement of the spleen in those responses. It was a time when physiologists were isolating and identifying new hormones and other naturally occurring substances in great numbers. Ungar began to characterize two protein fractions, spleenin A and spleenin B, that might play a role in the body's response to trauma. He also obtained experimental evidence that a host of different agents that could irritate or injure cells caused an increase in proteolytic enzyme activity. Proteolytic enzymes cause proteins to break apart. Aiding him in this work was Evelyn Damgaard, a research assistant eventually assigned exclusively to him. She would turn out to be one of the very few women that he ever worked well with.

Alberte, in the meantime, had remained with her sister in France, caring for her infant daughter while Ungar got settled. He sent for them in March, 1948, and they arrived in Chicago just as the northern Midwest was digging out of the paralyzing snowstorm that Ungar referred to above. To their dismay, housing in Chicago for a family with a baby could not be found. "Black people, Jews, and people with small children were definitely not welcome," Alberte remembered decades later. So befitting the culture of their newly adopted land, they joined the suburban migration—a forced decision with unfortunate consequences. For cultured and urbane Europeans, widely traveled and used to living in national capitals, the new suburban housing development of Park Forest seemed a sterile outpost at the periphery of their notion of civilization. Ungar joined a car pool, but couldn't stand it; whether because of his companions or the time wasted in commuting to work is not clear. For four years they endured Park Forest, but in 1952 found a nice place to live in Evanston, and moved immediately to the more congenial setting of that university community.

Ungar was unhappy for another reason. Though there were other senior workers at the Rheumatic Institute, no one of his caliber shared his interests, so he had no one to talk to about his research. The scientist who can produce in an intellectual vacuum is extremely rare; ideas grow out of conversations with those who thrive on like-minded esoterica. The most creative scientists depend on almost constant conversation. Lacking that necessary interaction with peers, Ungar found himself bogged down. In the context of that frustration, a controversy arose between him and the director of the institute, Alvin Coburn. The nature of the dispute is remembered differently by different persons, and Ungar was never clear about it, but it impugned his honor so much that nothing less than a formal retraction was acceptable. Northwestern University, the Institute's parent organization, fully exonerated Ungar in the affair, whatever it was. But apparently many of his colleagues felt he should have yielded to what they saw as a normal exercise of power, and this he interpreted as a betrayal. He was a stubborn man, unwilling to submit to what he viewed as bullying in the tenor of the McCarthyesque times, "one of the least attractive traits of the American character," in his judgment.

There followed a year at the University of Illinois, where Ungar got embroiled in another controversy—the famous dispute over "crebiozen," an alleged miracle cancer cure of the type that comes along from time to time. The cumulative frustration of professional turmoil and Chicago's lack of European charm left him anxious to return to Paris, when an opportunity to do so apparently arose in 1953. So anxious to leave Chicago were the Ungars that they decided to wait in New York "as tourists" for the lab in Paris to be set up. While there, the vice president for research at the U.S. Vitamin Corporation, an acquaintance of Ungar's, prevailed on him to help set up a product testing lab. In March, 1954, Ungar learned that the new French government had scuttled plans for the lab he was to head in Paris. U.S. Vitamin liked the temporary job he had done for them so much that they asked him to stay as the permanent director of the lab he had set up. The allure of New York was too much to turn down, so the Ungars settled into an apartment at 105th Street and Riverside Drive, and began what would be, until Houston, the most pleasant and productive period of Ungar's career.

From 1954 to 1962, he worked for the first time in his life for industry. His lab was well equipped and supplied, and he was given relative freedom to conduct research and publish as he pleased. Aside from the benefits that the product testing lab brought to the company, Ungar really earned his keep when he discovered, in collaboration with Seymour Shapiro, a brilliant young organic chemist killed by cancer too early in his career, an effective oral anti-diabetic drug—phenformin. It was widely prescribed for many years for diabetics who could not take insulin, and must have made millions for the company.

Ungar's first love and primary interest remained in basic research, which continued to focus on tissue response to injury. But he was beginning to look at the problem in a more general way; to look for common elements in disparate processes. He was struck, for example, by the prevalence of proteolysis—the tendency for proteins to break down—during the early phases of a tissue's response to trauma. It was known that proteolytic enzymes are required to break apart proteins in living cells, but it was also known that up to a point before they break apart, proteins can change their three-dimensional shape, or conformation, in a reversible manner, much as a rubber toy can be twisted into different configurations but still revert to its original shape. Ungar's awareness of proteolysis in damaged tissue led him to wonder whether non-damaging stimuli would produce the same response to a milder degree, namely a reversible change in shape in the proteins of the cell before they reached the point of breaking apart. He was aware of work done in France and Italy around the turn of the century, and in Russia beginning before the Second World War, showing that the chemical properties of proteins in nerve cells are altered when the nerve has been subjected to prolonged stimulation. So he set out to demonstrate that the alterations were related to changes in shape ("transconformations") in the proteins of nerve cells, and that more vigorous or prolonged stimulation led to the actual breakdown of proteins in neural tissue.

His ideas were provocative and prescient in the mid 1950s, when almost all physiologists were viewing excitation as purely bioelectrical phenomena; so he stirred some excitement among older biophysicists and a small but growing group of scientists just starting to call themselves neurochemists. His methods, unfortunately, were not as elegant as his concepts. Hardly unique among scientists, and certainly not for the last time in his career, he had an idea that could not be adequately tested by the technological capabilities of the times. But he tried, and the fallout from this effort produced a credibility problem for him that he was never fully able to shake.

The deleterious consequences of this research were not obvious in 1956, when he began to publish his results on transconformation changes and proteolysis in stimulated neural tissue. Others tried to repeat his work—some with success, others not. As the decade ended, work at the U. S. Vitamin Corporation was going well, life in New York was invigorating, and he was feeling good enough to undertake a project that he later swore, incorrectly, that he would never undertake again—the writing of a book.

Ungar had written a couple of monographs on restricted topics before the war, but now he decided to write a major work on the phenomenon of excitation in its entirety, this having become the common denominator of his research thinking. By the time it was published in 1963, *Excitation* represented a monumental effort, remarkable in several respects. It was very broad in scope, citing over 1200 references on the subject from its historical and philosophical roots, through its many manifestations in living tissue, to its pathological aspects and implications. It was unorthodox in focusing, not on the bioelectrical facets that constituted the usual way of measuring and studying excitation, but on chemical aspects of the phenomenon that were little studied and hardly agreed upon by physiologists at the time. The book devoted inordinate attention to hypothetical mechanisms that were alternative to the explanation most commonly accepted (now, as then) for how excitatory changes come about. Ungar left no doubt that his unusual treatment of the subject was deliberate by suggesting in his preface that "a disproportionately large space has been allotted to unorthodox opinions at the expense of the 'official' position" because of "a personal taste for heterodoxy allied with a distrust for all 'Establishments,' scientific or otherwise. Besides, as C. P. Snow said, nobody 'in his right mind will spend time expounding the majority view. There are plenty that will do that.' "

It appears that the book had relatively little influence. It did not sell enough copies to pay any royalty, and it has not been cited very much. Ungar blamed this on the publisher's failure to advertise the book or market it effectively. While probably true, it can also be said that the book's strength was its weakness—it was so comprehensive that it lacked depth at critical places, and it was uncomfortably integrative for the reductionistic predilections of most scientists. While it turned out to anticipate some notions that are more acceptable today, such as conformational changes in proteins associated with excitation, at the time it elevated the work and views of a distinct minority over the perspective of the vast majority.

Ungar had never expected to stay in industry, and as time went on in New York, he missed the academic affiliations that he had enjoyed in the past. He was offered almost simultaneously in 1962 the opportunity to become program director at the National Science Foundation, or to head an institute of comparative biology that the NSF was anxious to underwrite at the San Diego Zoo. With some hesitation, he decided to take the job in San Diego. It would cost him a big cut in salary. Also, it would be a difficult move for his daughter Catherine, especially, since she had grown up in New York and was in the middle of high school. (In fact, it proved too difficult; she opted to take her chances with college prematurely, rather than move to San Diego). But he was drawn to the tremendous possibilities for basic comparative animal research that he felt the zoo would provide. He would bring in a staff of senior investigators who, in affiliation with the new University of California at San Diego, would form a first-class collection of scholars. And he was admittedly drawn by the climate and scenery that promoted a migration of scientists to southern California of brain-drain proportions during the '60s. (When Samuel Barondes, a rising star at the Albert Einstein Medical School in the Bronx, moved to San Diego, one science reporter asked rhetorically, "How can you keep them in New York after they've seen La Jolla?").

With $60,000 to set up a lab, the Institute of Comparative Biology seemed to start out well in October of 1962; but it soon became obvious that the director of the zoo and his advisory committee had a totally different view of what the Institute should be. Ungar felt that his superiors wanted "just another exhibit," and he resented their resistance to academic affiliations of any sort. On top of this mismatch of goals, he and Alberte both perceived a climate of racial and religious bigotry that tarnished the weather and pleasant surroundings. He, at least, attributed this to the San Diego area in general, and spoke of it with some bitterness on more than one occasion in later years. So when his resignation was asked for in August 1963, he was glad to get out of the morass that this dream of a world-class research institute had turned into in less than a year.

Fortunately, Arthur Keats, a friend who had become chairman of the Department of Pharmacology at the Baylor College of Medicine in Houston, was able to offer Ungar a faculty position. With misgivings grown severe by this time after their unpleasant experiences in Chicago and San Diego, the Ungars moved to Houston. From what they had heard and read, they had gathered that "Texas had more of the worst than of the best features of the country." But they were pleasantly surprised to find the people of Houston more friendly and tolerant than those of San Diego, and the city decidedly less drab than Chicago. They found an apartment near the sprawling medical center, just blocks from the lab. It wasn't Paris, but it was a major city and an academic appointment with decent salary and good lab facilities. It would turn out to be the longest place they would stay in their married lives, and within just a few years would see Ungar come the closest he would get to the ultimate triumph of his career.

Catherine once told me that the relationship between her mother and father at home was tempestuous at times. "In everyday life he was a bit self-centered," Alberte confided. "He was so brilliant in everything. He knew so much about literature, music, and such things. I was in awe of him practically my whole life." That gave her an inferiority complex around him; but she was feisty and not lacking in gumption herself, so she tended to react assertively. "I would prove that I could do as well or better. That certainly created conflicts between us. He was not easy to live with."

But she remembers the nice touches as well. Occasionally, he would leave his study and seek her out, just to see what she was doing and say hello; then he would return to his work or reading. He liked having her at home, to a fault. "He was jealous of friendships—female friendships. Male friendships even more." This infuriated her at times, when he would complain about innocent conversations with male colleagues. But she also took it as the unintended compliment of an egocentric husband.

After settling in to their new apartment on Holcombe Boulevard in Houston, Alberte started getting stir-crazy. Catherine was gone, the apartment was small, and "I had absolutely nothing to do, so I said, Could I possibly work in the lab, and not just cleaning glassware?"

"But you don't know anything, anymore." he told her. Since the end of the war, she had been out of the lab, raising Catherine, making a home as he migrated from one job to another across the Atlantic, then across North America. It was true that she had been away from research for a long time and had not kept up with what he was doing.

"But can I try?" she asked.

"You're too old." It could not have been very flattering to her ego. Arthur Keats was on her side, however.

"Well, let's try her," the chairman advised.

So she came back to the research lab, and really did start from the ground up. She still remembers the day she picked up a pipette for the first time in 18 years and stared at the markings, dredging up the memory of what they meant.

She was put to work on the guinea pig ileum, the preparation that had gained him his first recognition and had brought them together. He really didn't think she could do anything, but she proved him wrong. Her steady hand, her knack for detail, and her patience brought back the memory of the old bioassay technique. "You have to have a certain feeling for that kind of work," she explained. As her confidence grew, she became the lab surgeon as well. She gained his grudging respect, then his confidence. Ultimately, he came to depend on her tremendously.

Not that they didn't have their frictions in the lab. She was practical, precise, and picky. He thought she was fussy and petty. He was theoretical, impractical, and impatient. She thought he was overbearing and too often unrealistic. But they both knew their places. "He was the brain, I was the brawn," she described it, and their characteristics were suited to their respective roles. In actuality, the metaphor was misleading, because he always liked to try procedures himself, before turning them over to others. And clearly there was intelligence required for the tasks that she performed. In time, the nice touches made their way into the lab as they did at home. He would wander over to her bench, to see how she was doing, to say hello. If she was out of sorts or concentrating on a critical pro-cedure, she thought he was nosy, and she was probably right. But other times she thought it was nice. They adapted to one another's foibles, and made an effective team for their remaining years together.

Opiates Point to the Transfer Strategy

The juice of the poppy plant has been known to elevate mood and inhibit pain from the dawn of recorded history. This "opium"—named after the Greek word for juice—consists of about 20 chemical compounds called alkaloids, the major one of which is morphine. Heroin is chemically similar to morphine and is two or three times stronger. Opium has been used since the time of the Greeks for medicinal purposes. But morphine and its relatives have toxic side effects. At high enough doses, they kill by causing respiratory arrest, and at lower doses they can produce nausea and constipation. By the eighteenth century, especially in the Orient, it was taken for recreational as well as medicinal purposes. In the nineteenth century, the influx of opium-smoking laborers from China, the liberal use of morphine to treat wounded Civil War soldiers, and the unrestricted availability of the drug promoted its widespread assimilation into the culture of North America.

The substance that William Osler referred to as "God's own medicine" has a sinister drawback. Like many drugs, repeated use of morphine or heroin requires a progressive elevation in the dose needed to elicit its beneficial effects. The body develops a "tolerance" for the drug, and needs more of it as time goes on. Not only that, but the body extracts a penalty by causing discomfort and illness if the drug is discontinued. This "withdrawal syndrome" reflects a physical dependence on the drug, presumably reflecting long-term physiological adjust-ments in the cells of the body. For some chronic users, the euphoric effects of the opiate drugs are great enough, and withdrawal symptoms are severe enough, to drive them to crime in order to obtain the substances illegally. The cost to the illicit user and to society is considerable. The motivation for scientists is high, therefore, to understand the mechanism by which these substances act, and to learn especially how and why tolerance to drugs develops.

When Ungar moved to Houston in 1963 he continued for some time the work begun in New York on conformational changes in protein related to function. Though the idea of such changes was gaining scientific respectability, Ungar's original methodology was still in question and he himself considered it unsatisfactory, so he wanted to clean up his credibility, as it were. But his interest in pain and analgesia, which had begun as far back as his work in the '30s with Tinel, was reactivated, perhaps due to the presence of William Krivoy on the faculty of the medical school. Krivoy was interested in the interaction between peptide hormones and drugs, including the opiates. A few years earlier, researchers in Yugoslavia had shown that a particular peptide called Substance P could diminish the analgesic effects of morphine in mice, and other examples of peptide-drug interactions were beginning to appear. So Krivoy and Ungar began to wonder if blocking the cell's ability to produce peptides would alter the effect of the drug on the cell.

The effect they chose to look at was the ability of morphine to induce "tolerance" in animals that were given the drug repeatedly. One of Krivoy's students, M. Cohen, performed the experiment by injecting mice with increasing doses of morphine every day. Some of the mice were also injected with actinomycin D—a drug known to block the formation of RNA, upon which the synthesis of proteins and peptides depends. All the mice were pinched in the tail daily, and those that responded to the pain were considered to be "tolerant" to morphine. The percentage of mice reacting to pain predictably rose each day as tolerance to the morphine injections developed. In mice given actinomycin D as well, however, tolerance developed to a significantly lesser degree. Control experiments showed that actinomycin D itself did not affect pain perception, so the logical conclusion was that by blocking the production of new RNA and/or new protein synthesis, they had reduced the ability of the cells to develop tolerance for morphine.

Cohen came to work in Ungar's lab, where he successfully repeated the experiment, and published it with Keats, Krivoy, and Ungar in June 1965. At the end of this experiment, there were morphine-tolerant rats left over, so Ungar and Cohen decided to extend the logic of the earlier experiment one step further. If the production of new protein is necessary for development of drug tolerance, animals that have already become tolerant must have the proteins in their cells. The presence of these proteins would be indicated if extracts of the cells containing them would transfer morphine tolerance into mice not previously exposed to morphine. Just as 30 years earlier he had tested for the presence of histamine in cell extracts by seeing if the extracts caused the guinea pig ileum to contract, he would test for the presence of tolerance-inducing substances by injecting an extract of tissue from tolerant animals into intolerant recipients. The experiment worked. Thirty-one percent of mice injected only with a salt solution were morphine-tolerant, but 64 percent of those injected with brain extracts from morphine-tolerant donors were themselves tolerant to morphine. The effect was not seen if extracts were made from any organ other than the brain. The central nervous system, it appeared, did indeed produce chemical substances that could transfer a physiological effect from one animal to another.

From the time that drugs and hormones first began to be studied, two notions about them had developed—vaguely at first, but with growing clarity by the early 1960s. The first was that a drug or a hormone doesn't make a cell do anything that it isn't ordinarily capable of doing, though it may cause it to carry out the ordinary function to excess, or shut it down altogether. The second notion was that to be responsive to a drug or hormone, a cell must have a molecular receptor—a specific site of action for that specific drug or hormone, just as a house must somewhere have a keyhole that accepts a key of a specific shape that can open the door of that particular house. In this analogy, hormones are the keys that trigger a cell to accentuate (or minimize) its normal function. A corollary of this logic is that a drug acts by mimicking or interfering with a natural compound, normally present or capable of being produced somewhere in the body, that acts at a specific site on specific cells. In retrospect, Ungar may have been close—very close—to discovering one or more natural compounds that act on whatever cellular site and system morphine acts upon. As a classically trained physiologist and pharmacologist, he must have been aware of these notions about drugs and their receptors; but at the time he was concentrating on the problem from another angle.

In *Excitation*, Ungar betrays a fascination with information theory, that constellation of ideas concerned with analyzing in quantitative ways the 'information content' of structures, systems, or machines. Spawned by the practical needs of the Second World War (where, for instance, survival depended on building weapons to process information for hitting a moving target), and fed by the growth of computer science, information theory was very much in vogue by the 1960s. With the exciting discoveries in molecular genetics begun a little more than a decade earlier, biologists too were focusing on the information content of molecules. Ungar, naturally, had been thinking along these lines since his work on conformational changes in protein (a key concept emerging in biology was the notion that shape carries information). In his mind, then, when he transferred morphine tolerance, he transferred information that the nervous system had encoded in chemical form in response to an external stimulus (the drug morphine). If the stimulus could be a drug, why couldn't it be any other information that made its way into the brain? Why wouldn't the brain produce a molecule in response to information about the organism's experience? And if it did, why couldn't it be said that this molecule represented the storage form of the experience? In short, why wouldn't it be a memory molecule?

Of all the forms of learning that Ungar knew about, habituation seemed the one most similar to the phenomenon of morphine tolerance. Habituation is the disappearance of a characteristic response upon repeated exposure to the stimulus that elicits the response (just as sensitivity to morphine diminishes with repeated exposure to the drug). If a horn sounds just as you walk by a parked car on a quiet street you may jump, but you don't jump each time a horn honks after hearing many horns on a busy street. Your nervous system stops issuing commands to "jump" after it has received repetitive information of no consequence. Since your behavior (jumping) has been altered by experience (hearing horns honk), you have learned something, in an elementary sense, and your nervous system must have a stored record (memory) of this behavioral alteration. If the record (memory) has been stored in a chemical form, transfer of the chemical to a naive subject ought to transfer the memory of the experience, provided the recipient can decipher the molecular code.

Ungar proceeded to expose rats to the clanging of an automated hammer every five seconds for two hours every day until they showed a startle response less than 10% of the time. Extracts were made of the brains of these animals and injected into mice; then the speed with which mice learned to ignore the sound of the hammer was recorded. Those mice given brain extracts from habituated rats responded less than 30% of the time, while mice given extracts from non-habituated rats responded about 95% of the time on the first day after injection. Every day for two weeks after injection, the experimental mice given extracts from habituated rats responded much less frequently to the hammer sound than did the control mice injected with non-habituated brain extracts. The experimentals, in other words, were acquiring the new behavior pattern strikingly faster than the controls. They were doing so, Ungar speculated, because the brain extracts were transferring the molecular storage form of the memory laid down in the brains of the donors. And the way the extracts were prepared, the molecules carrying the information had to be small proteins or peptides.

Summer of Revelations

I had stayed in Lubbock the summer of 1965 to complete the requirements for my undergraduate degree. With required courses in English Literature and Texas History still hanging over my head, I enrolled in early morning sections in order to finish my homework by noon, and work in the afternoons to save money for graduate school. This left the evenings free for work in the library, where I now poured over the growing literature on learning and memory that Hydén, Rosenzweig, Krech, Bennett and others were churning out. In particular that summer, I concentrated on a series of illuminating experiments by one of Rosenzweig's students, James McGaugh.

James L. McGaugh was born the son of a minister in Long Beach, California, in 1931. After high school in Riverside, he entered San Jose State University to study music and drama. In an introductory course in psychology one day he looked at a picture of a neuron and thought to himself, "If we knew how that cell worked, then my memorization in drama and music would be a lot easier." By his sophomore year, he was majoring in psychology. Moving to Berkeley for his graduate work, he completed his doctoral research with Krech and Rosenzweig. Daniel Bovet, the Nobel laureate whose discovery of anti-histamines had been aided by some of Ungar's early work, became McGaugh's postdoctoral adviser in Rome, providing the setting that would stimulate a lasting interest in the pharmacology of learning and memory.

Back in the United States, through a succession of faculty appointments at San Jose State, the University of Oregon, and finally, the University of California at Irvine, he initiated a series of experiments in which either stimulants or depressants were injected into rats after training. Inevitably, stimulants facilitated memory while depressants interfered with or depressed it. Clearly, the encoding of memory was not an instantaneous process, but rather depended on some type of progressive mechanism tuned to the general functional state of the brain. It was the best evidence then available that memory has an underlying metabolic basis. Taken together with the work on anti-metabolites by Barondes,

Flexner, and Agranoff, it seemed that learning depended on the synthesis of new protein, initiated by processes set in motion by immediate experience but not completed until a molecular form of storage for the long term memory was established.

So the scientific world was poised in the summer of 1965 for the discovery of a molecule, or set of molecules, associated with the long term encoding of behavioral experience. And I by then had read enough to be similarly poised myself. Thus on a hot day in July, I opened the latest issue of *Science* to read an electrifying report from the lab of Allan Jacobson, James McConnell's former student, by then at UCLA. Jacobson had operantly conditioned rats to approach a food cup, extracted RNA from the brains of the conditioned rats, and injected the RNA into untrained recipients who learned to approach the food cup on cue significantly faster than controls. The sample size was small and the rate of learning by the experimentals only marginally faster than the controls. It was the audacity of the experiment, as much as the actual result that struck me. If substantiated, this line of research was on track toward a discovery of monumental importance.

The summer ended, and I crossed the Red River into Oklahoma on September 1 with relief en route to Kansas, a new life, and the long-anticipated adventure of graduate school. I found an apartment on Tennessee Street at the foot of Mt. Oread below the University, and settled into a routine of study that included a continuation of my periodic review of the literature on learning and memory. It was in the course of routinely skimming the latest journals that I came across the report in *Nature* from the lab of Georges Ungar at the Baylor College of Medicine concerning the transfer of sound habituation in mice by injection of peptide extracts from trained donors.

When in the second week of September, 1965, I read the report of transfer of sound habituation by Ungar and Oceguera-Navarro, I sent a reprint request right away. The document, signed "With compliments, G. Ungar," arrived by return mail. Impressed with this personal courtesy, I began to wonder if a personal contact with such a scientist at the forefront of research on the biological basis of memory might be cultivated.

As Christmas approached, Dorothy Haecker, a friend and fellow student in philosophy from my home town of San Antonio decided that she wanted to go to a conference in Houston. The possibility of visiting Ungar's lab crossed my mind. So Dorothy and I determined to detour through Houston on our way back to San Antonio that first Christmas of our first year in graduate school.

To call Georges Ungar on the phone was an act of considerable brashness for me at that time. I would have much preferred to set up an appointment in writing, but I had sense enough to seize the moment, and was rewarded by a congenial voice and an invitation to come by his lab the following morning. We began our meeting by talking about the biology of memory for an hour or more. It was an animated conversation that fueled our enthusiasm and engendered mutual respect, as only such a discussion by two strangers who discover in one another a passionate and knowledgeable interest in the same subject can do. It was then, as we toured his lab, that he showed me the test tube of brain extract that had fired my imagination and imprinted into my memory an excitement that I have felt in science only two or three times since that day.

There was no question that I would give anything to work, if just for a season, on the search for memory molecules in Ungar's lab. I asked if summer research fellowships were available at the Baylor College of Medicine. He said that they well might be, and encouraged me to apply. As Dorothy and I left that evening, I was beside myself with excitement over the possibility of working in Houston during the forthcoming summer. I saw it, first and foremost, as a chance to link myself with a scientist at the cutting edge of a tremendously important area of research. But I was also drawn by the excitement of Houston's gritty gleaming sprawl across the bayou, as well as a frank desire to get out of Lawrence, especially since Samson wasn't going to be there anyway. Dorothy was equally wound-up over the philosophy conference she had just attended, so we talked at one another non-stop for the entire two hundred-mile drive to San Antonio.

My first semester in graduate school ended in January on a positive note. I finished with the highest grade in my physiology course, and was the top student in the lab portion of my experimental psychology course. Once finals were over, I threw myself into analyzing the possibilities for research with Ungar, and by mid February was able to send him a three-page commentary on biochemical studies of memory mechanisms with a list of suggested experiments.

Though tremendously excited by the possibility of working with Ungar, I think it's fair to say that I retained some skepticism. In a letter to him, I summarized the issues with appropriate caution:

> If biochemical transfer . . . is widespread, specific, and persistent, then engram fixation presumably resides in the synthesis and maintenance of specific molecules. If biochemical transfer is not widespread, is general, and does not persist as long as the response is retained, then apparently biochemical activity establishes or accompanies engram fixation, which then is permanent in the absence of the original biochemical conditions.

As I figured, Ungar had already addressed the ubiquity and specificity issues. Not only had he shown, to his satisfaction, transfer of morphine tolerance and sound habituation, but he was about to report that animals habituated to sound and those habituated to a brief puff of air both produced transfer factors which were specific to the method of training. So he wrote back, recommending that I work on the issue of persistence, to nail down the time course of the appearance (and possible disappearance) of the transfer factors. This was fine with me, so I settled down to surviving another semester of course work, rejuvenated by the prospects of the upcoming summer.

Ungar-to-Irwin-to-Samson

A formal offer of appointment as a research trainee at $400 per month—a phenomenal sum to me then—arrived from Ungar in late February. My unbounded enthusiasm was not tempered by Ungar's routine request for a statement from my Department that the work would further my graduate training. But Dr. Balfour, acting as my advisor in Samson's absence, did not share my enthusiasm and did not consider the request routine. Since the University had managed, only with apparent difficulty, to come up with a half-time teaching assistantship for me, I thought I was doing it a favor by going elsewhere during the summer; so I expected Balfour to write the waiver despite his grumbling.

If we knew how seldom anything is simple when we start our careers, we would never have the courage to try to become anything. Getting the waiver to go to Houston was not going to be simple. On March 2, Balfour called me into his office and read me parts of a letter from Samson instructing him not to let me go to Houston because, first, Ungar was a fraud and, second, Ungar would try to get me to stay there.

Notwithstanding the implication of my value to the Department, which I guess should have flattered me, I was shocked that my freedom to seek summer employment away from a home base that had not been overly generous with financial aid was being curtailed. And ignorant of the past unpleasantness between Ungar and Samson, I was totally confused by the charge of fraud; particularly at that time when fraud in science was regarded as virtually unthinkable.

But I felt a loyalty to Samson and Balfour that I didn't take lightly. They had given me a chance when others were unwilling to, and I wanted badly to "belong" at Kansas. So it posed the first real ethical dilemma of my scientific career.

I thought about it for 24 hours and decided, following the lead of then President Johnson, to ask Balfour if we could "reason together" on the matter. I laid out my logical argument, which Balfour listened to with hardly a word. When I finished, he sighed, said he had faith in my judgment, and admitted that on the whole the experience would probably be good for me. Whether he meant it, or simply saw that he had a mutinous graduate student on his hands who would not be deterred, he never said; but he agreed to write the letter of waiver. For that one decision I always respected him; not because he agreed with me, but because he was willing to disagree with Samson's directive and overturn it on his own initiative, a risk he clearly did not have to take. My part of the bargain was to write to Samson and explain the situation from my point of view.

The previous September I had written to Samson, inviting him to correspond with me concerning his activities at NRP, but he had not responded. I knew the letter I was about to write would get a response. If Ungar is a fraud, I asked, how can you tell without firsthand observation? If his ideas are nonsense, where better to debunk them than in his own lab? If he is unethical, would I not have the intelligence to see it or the integrity to resist it? As for staying in Houston, this was out of the question because (1) I didn't want a degree from a medical school, (2) I didn't want a degree from another school in Texas, and (3) I liked the University of Kansas. As for controversy, I was still young enough to discount the downside that it can have on a scientific career:

> I realize that Dr. Ungar's work is controversial. In fact, I am quite skeptical myself of the generality of his conclusions, and my proposed project, which he approved, is designed to test one of the basic aspects of his theory. It is not because his work is controversial that I want to go; neither do I want to refrain from going only because it is controversial.

A letter from Samson came back by return mail. It was a pleasure to have my "excellently reasoned" letter, he wrote. "Although I feel you will be disappointed when you see Dr. Ungar's approach to data collection and will wonder if one should spend a valuable summer finding that out, I can not answer the power of your arguments. You have thought it all out so I can't help but agree with you."

He, of course, had thought it out too. He may have bought my argument that I might be able to prove that Ungar was the fraud he suspected him of being. Or it may have been that I represented a pipeline directly from Ungar's lab back to NRP. On the off chance that Ungar turned out to be right, he would be among the first to know, and that couldn't hurt him at NRP. Besides, it was not in his nature to be close minded, even about his adversaries. Whether Frank Schmitt knew or cared if Samson had a student who was about to work with Ungar, I don't know. What Ungar was talking about was not inherently contrary to notions that Schmitt had entertained, and had he known I was going there, a Samson-to-Irwin-to-Ungar connection would have been in Schmitt's interest to encourage. But this is all speculation. The bottom line was that Samson gave his approval, if not his blessing, and my summer adventure was back on track.

The semester ground on toward the summer, not nearly fast enough for me. In April, Balfour informed me that Samson had decided to stay another year at NRP. This depressed me, because it meant that another year would go by before I would get to solidify the type of relationship with my graduate advisor that I

had hoped for. Any guilt feelings I had about going to Houston certainly evapo-
rated at that point, and I began to think seriously about getting a master's degree
at Kansas and going somewhere else, though I intended to stick to my resolve
not to stay at Baylor.

At long last, the semester ended and I headed south once more. My first
Spring in Kansas had been cooler than I was used to, and the warmth felt good
as I approached Houston in the early morning hours of June 3. I still remember
the stars over Texas and the warm humid air that rose from the swamps as the
Gulf coast came nearer that night. Many people find the climate of Houston
oppressive, and even I was to suffer from the heat before that summer was out,
but most of the time I liked the way the sultry atmosphere of the bayou envel-
oped me and, that summer at least, brought me comfort.

Within a week of my arrival, I was hard at work on my own project. Rats
would be trained to go either to the left or the right in a Y-maze. Most rats
learned the proper choice in only a few days of training. Elaboration of the transfer factor in the brain of the donors was tested by making an extract of small
molecules including peptides from the brains of the donors and injecting the
extract into the abdominal cavity of recipient mice. If the transfer factor were
present, presumably it would be picked up by the blood stream and transported
to the nervous system where it would influence the recipients to turn in the
"correct" direction. My first experiment was set up to train rats for 3, 6, 9, 12,
15, and 18 days. Since there were six rats in each group, I started training 36
rats. Each was given 20 trials a day, so this meant my days were long. Running
rats after the first few thousand trials lost its novelty, but I went home every
night with a sense of accomplishment—certain that I was working on one of the
most intriguing and important problems in all of science.

By the time I got to Houston that summer, though, a backlash against the
initial transfer experiments had set in. Less than five months after *Science* had
published the remarkable experiment by Jacobson's group at UCLA showing
transfer of food cup approach behavior, the journal published a short note by
Gross and Carey reporting an inability to reproduce the results of Jacobson,
though their yield of RNA was much lower. In February, *Science* published a
longer article by James McGaugh and co-workers at the University of California
-Irvine reporting an attempt to transfer over half a dozen different types of
behavior by RNA extracts, none of which were successful. McGaugh's group
further showed that very little RNA injected into recipient animals actually
reached the brain. Even though none of the experiments replicated the conditions that Jacobson had used, the report was particularly damaging because of
McGaugh's stature in physiological psychology at the time.

These reports were only the tip of the iceberg. Ungar showed me a file in
mid-June that contained a huge pile of informally circulating manuscripts, most
of them reporting unsuccessful attempts to replicate transfer of the learning tasks
reported by Jacobson with RNA extracts. I well remember the sense of excitement I had in reading through that folder, realizing that the work I was doing

was right at the forefront of what the biggest names in psychobiology were grappling with. The failure to replicate the RNA transfer experiments didn't disturb me much, because I thought it extremely unlikely that RNA acted as a memory molecule in the way that McConnell had envisioned. No one had reported a failure to transfer sound habituation by peptide extracts, as Ungar had apparently demonstrated. Furthermore, on New Year's Day of 1966 Nature had published a replication of Jacobson's experiment by Frank Rosenblatt and co-workers at Cornell University that had given positive results, with a difference. Rosenblatt incubated his RNA extracts with an enzyme known to break down RNA, but the transfer effect remained. This suggested that something other than RNA—proteins or peptides, perhaps—were getting into the RNA extracts and actually causing the transfer effect.

It began to fit together. Nobody could replicate the transfer effect with RNA. Those who reported positive effects were attributing their results to pro-teins or peptides. The original reports of successful transfer with RNA must have been due to contaminating proteins or peptides in the extract. Ungar and I, and now Rosenblatt, were on the right track. All I had to do to nail it down was show whether the presence of the peptide coincided with the appearance and loss of the memory.

Day after weary day I spent in the rat room. As I finished each donor group, I had to kill all the rats, dissect out their brains, and make the peptide extracts, on top of continuing to train the remaining donors. I shuttled back and forth between the rat room and the biochemistry lab, in 13-hour and 14-hour days that began to run together. My desk was in an office at the other end of the building, and I had to pass through the complex of labs where Michael Debakey was implanting artificial hearts in calves. Downstairs, Roger Guillemin was working on peptides (thanks to a suggestion from Ungar) that cause the release of hor-mones from the pituitary gland. It was a heady and stimulating atmosphere. I went home every night, and often late at night, bone tired but satisfied.

When I had the energy, I wrote letters. I felt that it was in my best interest to keep Balfour and Samson informed, more to reassure them that I was able to keep my rationality and integrity intact while in Houston, than to pass along information. But I also knew that Samson would appreciate the information. In my first letter to Samson from Houston, I was still hedging my bets, but I had seen enough to convince me that under certain circumstances, the transfer effect was real and not due to any irregularities:

> I have looked at data on all these experiments—much of it the raw data that he and the technicians collected—and have watched them go through the various training, testing, and biochemical procedures. Although I want to suspend final judgment until I run my own rats and do my own calculations, there is not much doubt in my mind that he is looking at a real phenomenon here. I know it seems a little fantastic . . . but the data keep showing this.

I further pointed out that all the information on the transfer factors contin-ued to point to small peptides or proteins. For years it had been known that two important hormones, oxytocin (which causes uterine muscle to contract and breast milk to be ejected) and vasopressin (which promotes water retention), were small peptides manufactured in the brain. So I confessed in a marginal note in my letter to Samson that I worried that when the transfer factors were purified, they would turn out to be oxytocin or vasopressin. Years later, after many more peptides were discovered in the brain, many of them including vasopressin were indeed shown to have an influence on behavior—some of them rather specific behaviors. About that time, Bela Bohus and David De Wied in the Netherlands were turning my "worry" into a real experiment. They demonstrated that adrenocorticotropin, a large peptide from the pituitary gland that activates the adrenal gland, speeds the rate at which rats can learn certain tasks. At first, they were not believed, but De Wied persisted with his experiments and in time came to be recognized as a pioneer in the field of peptides and behavior by younger scientists who never heard of Georges Ungar. In the summer of 1966, we were not interested in non-specific effects of known hormones. We were gambling on a more spectacular discovery.

Samson responded to my letter by return mail. He was especially eager to hear of my progress "because there is so much controversy and disagreement about transfer experiments." He added, "Dr. Ungar seems to be almost alone in some of his conclusions." He wasn't ready to buy into the transfer experiments yet, but he was obviously anxious to pursue the dialog with me, and this made me very happy. I could see the possibility that we might develop a good advisor-student relationship after all. I had worked for 16½ hours that day. I went to bed that night more tired but more satisfied than usual.

The results of my first experiment began to come in, one group at a time. I plotted the percent of left and right turns for each recipient, depending on whether the mouse had been injected with peptide extracts from left-trained or right-trained donors. The results were not overwhelming. On a hunch I decided to group the data for 3, 6, and 9 days of training, and the results of that were staggering. For four days prior to injection, mice hovered around 40% turns to the left (60% to the right), but by the fourth day after injection, recipients of left trained brain extracts had risen from 38% to 52% left turns, while mice receiving right trained brain extract had gone from 59% to 71% turns to the right. The possibility that the mice could have changed their behavior that much by chance alone was less than 2 in 1000. This time I had run the rats, and I had done the calculations. Something that looked like memory transfer was real.

In terms of the original goals of the experiment, though, the results were confusing. Analysis of the different donor groups according to length of training showed that the best results had come from the donors trained the shortest length of time. When we looked at donor groups trained for 12, 15, or 18 days, we observed an increase in percentage of turns to the left, whether the recipients were injected with left-trained or right-trained donor brain extracts. In other words, the left-turning tendency transferred as expected, but the right-turning tendency transferred in reverse. It was a strained explanation, but it made sense to us at the time.

It began to look like the phenomenon was complicated. There appeared to be two factors, all right, but they didn't rise consistently nor did they change together over different lengths of training, and they apparently could cause the recipient to behave in the opposite as well as in the same way as the donor. But I no longer questioned that there was something in those brain extracts capable of influencing behavior. The picture of that first graph was etched in my mind. The data were there. Unfortunately, I would never get data that good in a transfer experiment again.

As the summer progressed, I became a friend with the Ungars' daughter, Catherine. She was majoring in anthropology, and since I had dabbled in the subject during my undergraduate years, we found ourselves arguing theories of human origin from time to time. These spirited discussions inevitably broadened to the current topics of the day—civil rights, the war in Viet Nam, the space race, the battle of the sexes, the relevance of a liberal education. I was going through a cynical phase at the time, but her wit, freshness, and irreverence gave me a sense of perspective that I needed.

Just as my first and apparently successful transfer experiment was ending, the Ungars had dinner for the whole lab at their apartment. I relished the opportunity to strengthen my social bonds to all the Ungars as well as other members of the lab; and I certainly welcomed the break from the tedium of running rats and mice all day. The Ungars, though low-key in the lab were all business there; in their own home they really relaxed, turned on their European grace and erudition, and laced it with Texas-style hospitality. It was a devastating blend for me, and I felt as close to them and that group by the end of the evening as I had to anyone in a long time. Houston was beginning to feel like home.

But I had meant it when I had told Samson I wouldn't stay there, and I continued to keep him informed. I knew that Samson would be convinced by the cold facts, once they were sufficient and compelling. A man who had been an acrobat and become a professor—who had converted from a dedicated socialist to an unabashed free-enterprise conservative—was not going to fly in the face of reality forever. The extinction experiment would convince him before the summer was out.

Extinction is psychological jargon for "forgetting," as applied to animals that demonstrate their loss of memory by failing to perform a previously learned task. In my experiment, I would train rats, then stop running them long enough for them to forget the proper left-right choice. Others would be trained, then left alone for some time but not long enough to forget the correct choice. Those rats for whom the memory had become "extinguished" should not produce transfer factors for a non-existent memory, if memories and molecules were materially matched to one another. By the same token, transfer factors should be present as long as the memory persisted in donors who remembered the correct way to turn in the maze.

The experiment was outlined. The schedule was set. The rats and mice were ordered. But on July 8, 1966, the International Association of Machinists struck five major airlines, and my beautiful experiment—the key to my professional future—couldn't get off the ground. We were absolutely dependent on an animal supplier in New Jersey for our consistently healthy supply of rodents, but nothing that wasn't human or human luggage could get on a domestic flight for any price for 43 days. Jetliners sat on the ground across the land. My rats and mice grew older in their cages in New Jersey, ignorant of the scientific break-through to which they would never contribute.

Tragedy and Speculation

Charles Whitman, an engineering student at the University of Texas in Austin with a B average and seemingly pleasant personality, got up in the early morning hours of August 1 that summer, drove to his mother's house and killed her, then returned home and stabbed his young wife to death in her bed. During the rest of the morning, he assembled an armory of weapons and provisions, packed them into a trunk, and took them up the elevator to the 27th floor of the library tower at the center of the campus. There he beat and shot to death the receptionist at the entrance to the observation deck, and shot a family of tourists. Just before noon, he moved his trunk with weapons and provisions out onto the deck, and proceeded to fire at random human targets on the grounds and streets below. Within 20 minutes he had killed a dozen people. An Austin policeman and high school friend of mine, Bill Speed, was one of them. Whitman withstood the siege for an eternity, protected from ground fire by a limestone parapet that ringed the observation deck. When it was clear that nothing else would work, two police officers and a deputized civilian climbed the stairs to the observation deck. With skill, some luck, and considerable courage, they finally brought the terror to an end by shooting the sniper before he could hit them.

It was time for me to write to Samson again. I wasn't in the mood to do it that night, but I felt the need to do something useful in the face of the useless tragedy that had killed a friend and defiled a treasured landmark in the state of my birth. I sat down that night to write a progress report, and to explain the best way I could how the transfer phenomenon could be incorporated into our understanding—so pitifully inadequate as the day's events had shown—of how the brain works.

I began by pointing to assumptions that everyone agreed upon: that any learning task involves millions of neurons broadly distributed throughout the brain, and probably requires a sequence of changes rather than a single change in a given molecule, or neuron, or neuronal connection. Then I pointed to a self-evident assumption: that in learning different behaviors (say left turns versus right turns), a different set of neurons is called into play.

At this point I advanced the hypothesis upon which the *specificity* of memory transfer absolutely depended: that different neurons act upon or produce different molecules in the course of their activity. The million (I'll use round numbers for simplicity) neurons that direct the body of the rat to turn left are different from the million neurons that direct it to turn right, and accordingly act upon or produce a different set of molecules.

Finally, I explained how the transfer of a behavioral response could occur: When the population of molecules from a left-trained brain found their way into the brain of an untrained recipient, they shifted the molecular milieu in the recipient's brain toward what it would be if the million neurons required for left-turn learning had been activated. If—and here I made another assumption—neurons are sensitized by the molecules that act upon them, the altered milieu will lower the threshold of activation for the million neurons that mediate left turns more readily than the million neurons that promote turns to the right.

It was a population view of brain function. It came out of my earliest training in biology, literally from the lizards of the desert, where the group had a dynamic that transcended its individual units. It was heavily influenced by the writings of Ross Adey and E. Roy John, with their emphasis on "patterns" of activity in populations of brain cells. It didn't transgress anything we knew about neural function, but it did implicitly reject the notion that there was a one-to-one correspondence between memories and molecules.

> I don't think we're transferring specific molecules 'coded' for certain behavior patterns; only that we're changing the functional biochemical environment, hence the function, of the central nervous system in a crudely predictable direction.

By August of 1966, I had realized that the notion of a memory molecule was overly simplistic. But that was no reason to doubt that the molecular milieu of the brain could influence the pattern of activity in its functional units. Today, that assumption is not even seriously questioned.

"Failure to Reproduce Results
Is Not Unusual"

If anyone knew or cared what was going on in Houston, the published literature gave no indication of it. Starting with the Gross and Carey report of failure to replicate Jacobson's transfer of food cup approach with RNA in December, a drumbeat of negative results had built through the winter and spring. Then on August 5, *Science* published an extraordinary report signed by 23 authors from eight different laboratories, announcing an inability to transfer a variety of tasks with RNA, including the one originally reported by Jacobson. The report was restrained—almost apologetic—and respectful of the approach.

> . . . it would be unfortunate if these negative findings were to be taken as a
> signal for abandoning the pursuit of a result of enormous potential significance
> . . . Failure to reproduce results is not, after all, unusual in the early phase of
> research when all relevant variables are as yet unspecified.

The effect of the report was exactly what it asked not to be. With 23 scientists representing some of the world's leading laboratories in neuroscience saying that they could not repeat Jacobson's experiment, there was no way that the transfer approach could avoid being irreversibly tainted. In many cases the experiments differed from what Jacobson had done, and in all cases it was only RNA extracts that had been tested. At that point, there were still no published refutations of Ungar's transfer of morphine tolerance or sound habituation by peptides, or of Rosenblatt's transfer of conditioned responses by non-RNA substances. But that distinction was lost in the wave of prestigious negative publicity. It was a crippling blow from which the transfer approach was never able to recover.

The senior author of that paper, William Byrne, was an unlikely and unwilling agent for the demise of the transfer approach. Ironically he was to become the paradigm's chief protagonist and facilitator in the years that followed. At the time, he was working in Melvin Calvin's lab at Berkeley with Ros-enzweig and Bennett, looking for a way to redirect his career in biochemistry toward the brain sciences. When the Ungar and Jacobson reports of transfer appeared, he suggested that the lab pursue them with vigor. It must have been galling for Rosenzweig and Bennett, who had worked for a decade already to prove some very marginal anatomical and chemical changes associated with environmental stimulation, to face up to the possibility that such an elegantly simple approach as memory transfer would unlock the secret of the neural mech-anism of learning. But to their credit they did face it, committing years of work and untold resources to an endeavor that ultimately produced very little.

Byrne's career had not been uneventful. A graduate of Stanford, he had earned his doctorate at the University of Wisconsin through research on enzymes that break down sugar to yield energy. After postdoctoral studies at the National Institutes of Health, he was offered a faculty position at Duke in 1954. Grants were not difficult for good researchers to obtain, and he built a sizeable research program that was among the first to generate a couple of important ideas in biochemistry. One was the notion that the end-product of a metabolic reaction could inhibit another metabolic reaction earlier in the sequence leading to production of the final product. To avoid the tremendous energy surges that would overheat living cells, most metabolic changes consist of a sequence of gradual modifications, like a series of small rapids rather than a single huge waterfall. The total volume of water downstream, like the total number of metabolic products at the end of a sequence of reactions, can be controlled by limiting flow through one or more of the small rapids (metabolic conversions) "upstream." Byrne and his graduate student, Frank Newhouse, had discovered that as the amino acid serine builds up, it inhibits one of the upstream reactions leading to its production.

Another important idea that came out of Byrne's lab was the notion that some enzymes are activated or inactivated by adding a small molecular cluster called a phosphate group (for the phosphorous atom at its center). It was another twist of fate that the phosphorylation reactions would become central to memory mechanisms postulated decades later by Eric Kandel, Gary Lynch, and others. But Byrne wasn't even working on the nervous system at Duke in the 1950s, much less thinking about memory mechanisms. By 1965, on the other hand, he felt the wind blowing in the direction of neuroscience, and he looked into a sabbatical with Frank Schmitt at NRP, and with Melvin Calvin at Berkeley. Calvin had done the type of solid biochemistry with which Byrne was familiar, and the Laboratory of Biodynamics had a record and reputation for being open-ended and innovative, as exemplified by Bennett's presence and his work with Rosenzweig, Krech, and Diamond for many years. So Byrne opted for Berkeley.

Byrne, Bennett, and their colleagues at Berkeley worked very hard trying to replicate Jacobson's finding. They obtained rats from the same supplier that

Jacobson had used, had Jacobson's technician come to Berkeley to extract the RNA, and had Jacobson himself watch the work. In spite of this attempt at exact replication, they could not repeat the results that Jacobson had obtained. Other labs across the country had mounted a similar effort. Except for Ungar and Rosenblatt, who were not focusing on RNA, the results were generally negative. As the evidence mounted that Jacobson's experiment was not easy to replicate, a decision was made to distil the multitude of unpublished manuscripts from the different labs into a short summary report for *Science*. Byrne felt that it was too early to publish negative results, given the vagaries of the experiments they were attempting. But the majority ruled and the note was submitted with all 23 signatures, bunched by institutional affiliation but not otherwise in alphabetical or any other discernible order. The Berkeley lab was the first one listed, and Byrne's name was the first within that group, though why his name came first (Bennett's was third) he never knew. Perhaps it was a courtesy to him as a visiting scientist. It was someone else's decision, in any event, but a decision that he let stand. For those old enough to remember the heyday of the transfer experiments, more often than not it is the short note by Byrne *et al.* that they cite as a reason for concluding that there was really nothing to the purported phenomenon. Very few know how badly Bill Byrne wanted the phenomenon to be real.

Welcome Back, Plain Planes—
You Ruined My Career

My anxiety grew as day after day of August rolled by with no end to the airline strike in sight. In desperation I finally ordered some rats from a local supplier, but half of them were sick and their performance was so erratic that I had to throw out the data from them in all good conscience.

Finally it ended the second week in August. Braniff Airways, which flaunted the uniqueness of its brightly colored fuselages, ran a full-page ad with a picture of a flight attendant in her outlandish Pucci outfit sitting on the floor rubbing her feet in pained exhaustion. The caption beneath recited the number of meals and cups of coffee and miles and airports and flights she had served through the strike, ending with the grateful exclamation, "Welcome back, plain planes!" Braniff and Continental and a couple of others had stayed in the air, but they couldn't get my rats from New Jersey to me. With the plain planes flying once more, I still thought I had a slim chance of running the experiment that would make or break the transfer phenomenon by the end of the summer.

It wasn't to be. My rats arrived on August 15, and I started training them immediately for the extinction experiment. After three days of training, I waited for seven days more, then tested their ability to recall which way they had been trained in the maze. I was pleased but not too surprised that they remembered; I had really wanted to wait longer. The question then was whether extracts from their brains would transfer memory of a behavior that had not been rehearsed for a week. The results were inconclusive at best. In a couple of the groups, the recipients showed a tendency to turn more in the direction of their donors, but not until four days after receiving injections of the donor extracts. By then it was the second of September and I didn't have time left to see if the trend continued.

While my future as a brain scientist hung in the balance those last two weeks in August, Ungar totally surprised me by asking me to co-author with him a chapter on chemical aspects of brain function that he had been invited to write for a forthcoming book. For a person at my level of training to co-author a review chapter with another senior scientist was (and still is) unusual. Whether he genuinely needed the help, or primarily wanted to give me a boost in the face of mounting discouragement with my experiments, I never knew.

I didn't accept immediately. I was wary, not only of the time it would take from my graduate studies, but by the suspicion I feared it might be met with from Balfour and Samson in Kansas. I would think about it, talk to Samson, and decide once I got back to Lawrence. But Ungar acted like he really hoped I would do it, and that did wonders for my morale at the end of an exciting summer that ended on a disappointing note.

Progress

"So You Found Someone Interested in This Heretical Idea?"

Back in Kansas and undaunted in September of 1966, I located a comfortable if creaky apartment for $65 a month, and plunged into the new semester. Damon Mountfort, a recent graduate student of Sebastian Grossman (author of far and away the best physiological psychology textbook available at the time), was offering a course on the Physiology of Motivation, which fit in perfectly with my desire to minor in psychology. Marjorie Newmark in my own department was offering her course in Cell Regulatory Mechanisms—the study of the various factors that control gene expression and enzyme activity in cells. Knowing the possibilities for controlling the metabolism of molecules in brain cells was clearly essential to getting a grip on alterations induced by experience. Already I realized that most of the speculations about molecular mechanisms of memory storage—and they were a dime a dozen—were ridiculously simplistic. With McConnell ignorant of biochemistry, and Ungar knowing little and caring less about behavior, I was determined to learn both subjects well.

The semester had just begun when the coordinator of the graduate student seminar asked me to talk about my summer's work seven days hence. Alarmed at the short lead-time but anxious to get it over with, I threw myself into preparation for this event, my first formal presentation before the faculty and students of my department. I would be speaking for the first time to the professors who controlled the next three years of my life, and I would be speaking on a topic already regarded with skepticism by much of the scientific community, and with deep suspicion by my chosen major professor (whose absence in this case—he was back in Boston at NRP—was just as well). So it was a big deal, to the point that I missed "Batman" several times during the ensuing days of hectic preparation.

My seminar presentation, "Biochemical Transfer of Learned Responses," was not perfect, but was good. The 40-minute talk was the longest I had ever given, and it went by in a flash. The audience was attentive, though fear was

obvious on the faces of many of my student colleagues who clearly were thinking forward to the morning when they would have to be in my place. I got the
impression that the faculty was a bit surprised. I had not done particularly well
in my intermediary metabolism course the previous spring, and my mediocre
undergraduate record had probably left some of the faculty skeptical about ad-
mitting me in the first place. My attitude and my earnestness had even been in
question, or so I perceived. So my obvious enthusiasm for this topic, and my
apparent instinct for research (which course grades seldom reflect, but research
presentations readily reveal) must have won them over, judging from feedback
in the ensuing days and the fact that my attitude was never again an issue.

A week after my seminar I wrote a long letter to Ungar, telling him how it
went and relating the feedback I had gotten. He responded by expressing his
pleasure that "you found people who are interested in this somewhat heretical
area . . ."

The transfer phenomenon in mammals had first been reported less than a
year and a half earlier, but a sense of paranoia had already set in. I think in
retrospect that the paranoia, or at least the need to aggressively defend the phenomenon, in the long run led its adherents down a futile and counterproductive
path. At the time, I was reasonably confident that truth and objectivity would
eventually triumph. I still believe they do; but the long run, I've learned, can
turn out to be a long time.

Ungar and I then began to correspond in earnest about the chapter on chemical correlates of neural function that he had proposed we write together. I was
very leery of getting involved in a project that would demand so much time, and
would have been just as happy if he had forgotten about it. But he hadn't, so I
allowed myself to be drawn in. With Samson still in Boston, I went to Balfour
for advice, and again he advised against further involvement with Ungar. His
argument, to be fair, was based on the time commitment, not on Ungar. History
was repeating itself. I braced for another long letter to Samson.

On November 1, I wrote him for the first time in three months, summarizing
the way the summer had ended in frustration for me in Houston, debriefing him
on my seminar, and asking him outright what he thought about my working on
the chapter with Ungar.

"I think you should accept the invitation to be a co-author on the chemical
correlates of neural function," he wrote back. "On the over-all it will be an interesting project and worthwhile."

Then he went on to put a finger on a characteristic of Ungar's critics that I
later came to recognize with consistency. "I am puzzled in some ways by Dr.
Ungar. He seems to be 'untouched' by the criticisms of others, and the fact that
people cannot repeat his experiments." There were still no published failures to
replicate Ungar's transfer experiments at that point, and it was beginning to
bother me that everyone was glossing over that point. The first part of the sentence was, of course, revealing. We all want to be taken seriously by our adversaries.

Samson next launched into a lengthy critique of the whole behavioral transfer field. He had heard Rosenblatt, Fjerdingstad, and Jacobson, and found them all unconvincing. Dave Albert, his nephew by marriage who had worked at Kansas for three years before moving to Canada, had tried some transfer experiments and "thinks something is there," but had apparently dropped the project, which made Samson suspicious since he had a lot of confidence in Albert's abili-ty. He repeated the charge that "a number of labs have tried to do the 'transfer' experiments and can not repeat them." But "I guess that can be accounted for, too," bringing him to the focal point of his dissatisfaction:

> There are so many variables involved that 'failures' to get expected results can always be accounted for. I am suspicious when new and more involved 'accounted fors' and 'biases' and 'overlookeds' keep coming in to language of the people claiming 'transfer' . . .

It was a Catch-22. The very explanation of why someone else cannot repeat a result is used as a reason to dismiss the claimed results in the first place. It didn't seem right, but somewhere deep in my scientific soul, I sensed the handwriting on the wall. I began to see a futile battle stretching endlessly before *me* and everyone else who felt that the transfer approach had promise at this point in the history of neuroscience. I don't honestly remember whether Samson triggered this insight, or confirmed a realization already growing in my own brain, but somewhere around the first or second week of November, I decided to concede.

I would not concede without protest, however. I pointed out in my answer that everyone failing to replicate the transfer effect was using nucleic acids rather than peptides; and that it wasn't unreasonable to expect numerous failures "when the phenomenon (learning) is so intricate and the experimental approach (transfer experiments) so gross." But I conceded the point that as long as the phenomenon remained marginal and unpredictable, there would be so many ways to explain any result that no result could have real meaning. Therefore, I concluded,

> That is why I have thought it better to take advantage of the strengths in our department, and attack the problem in a somewhat different way . . . In time, those who are well set up to do the transfer work, such as Dr Ungar, can probably offer more definitive and meaningful data.

In time Ungar did report transfer of behavior that was far from marginal, and went on to isolate a specific peptide that he thought was the mediator of the effect. But he wasn't believed by enough of the people who counted in the places that mattered. I watched the story unfold in fascination for years, but I watched from the sidelines, sadly correct in my assessment that it was not a winnable cause.

Samson was always open to disagreement and tolerant of offbeat ideas, almost to a fault; but in the end he had a good sense for what was scientifically effective and achievable. Unknown to me, he had decided to take a stand. Had I insisted on pursuing behavioral transfer research at Kansas, he was going to ask me to find another advisor. My own decision to try another approach saved him the need for disavowing me, so for better or worse, we stuck together for the rest of my years in Kansas.

Hebb Redux

On December 8, Samson arrived for a brief visit to work on the renewal of his long-standing NIH grant on the energetics of the nervous system. The energy level of the department began to perk up days in advance of his arrival, and no one looked forward to his coming more than me. For months we had engaged in a correspondence that was stimulating, but inordinately devoted to a challenge and defense of the transfer experiments generally and of Ungar in particular. Through it we had come to understand and respect one another's point of view, but it hadn't done a lot to clarify what I would do for the remainder of my graduate training. Writing an NSF Fellowship proposal forced me to focus my thoughts on the type of research that I thought was important and achievable, but it was all done without benefit of Samson's advice. I was very anxious to get his reaction. For that matter, I was anxious to talk to him face to face about anything, as by that point I had been in graduate school for 15 months and had yet to have a conversation with the person I had come to Kansas to work with.

On the morning of his second day in Lawrence, he and I were finally able to meet for an hour; then later he had me sit in on a meeting of the entire lab group. My NSF proposal had reached him just as he was leaving Boston, and he had read it on the plane. He was extremely complimentary about the proposal, to the point of thinking that one part of it ought to be incorporated into the resubmission of his NIH grant proposal. Other members of the group were not so enthusiastic. They thought it was a long shot for success, and did not think that it related sufficiently to the topic of neural energetics, to warrant inclusion.

That evening I assessed the net result of my first day in person with my doctoral advisor. On the one hand I was delighted by his positive reaction to my NSF proposal and the way he had defended an idea of mine to the whole lab group. Though I had been actively participating in the research meetings of the lab for months, this episode really made me feel that I was an integral and valued member of that group. On the other hand, I was a little dismayed by Samson's tendency to wander and speculate and think out loud with little evident focus and no obvious organization. I was particularly disappointed that we never got to some other parts of my proposal, or to a discussion of his future or mine. But that would come later, I presumed.

A short time before Samson's visit, I had gone to hear Andrew Nalbandov speak at Kansas State University on LH-RF, the LH releasing factor. LH stands for "luteinizing hormone," another one of the hormones released from the pituitary gland at the base of the brain. In humans, it promotes ovulation in women and testosterone secretion in men. Like most of the other pituitary hormones, its secretion is triggered by a releasing factor, produced in the hypothalamic area of the brain just above the pituitary. At that time, evidence was accumulating that each of the pituitary hormones had its own releasing factor, and that they were rather small molecules about the size of a short peptide. On December 1, I relayed to Ungar the information that Nalbandov thought the LH releasing factor was a small peptide, adding:

> This was interesting, since more evidence seems to be indicating that these releasing factors are short-chain polypeptides . . . I have been thinking about this with respect to the transfer experiments, which also implicate polypeptides . . . It may be a general result of nervous excitation that protein fragments, polypeptides, or individual amino acids are released by individual nerves. In some instances these become neurohormones . . . in others they become releasing factors . . . in still others they become specific neuronal markers, which may constitute the transfer factors.

I went on to ask him if he thought this might provide an orienting perspective for the chapter on chemical correlates of neural function that by now I was committed to write with him. He had thought of all this, and was quick to respond:

> What you say about releasing factors and neurohormones is very close to what I believe. The release of intracellular material is a very common feature of excitation . . . Out of this phenomenon evolved in the nervous system the neurosecretory process, the release of transmitters, of releasing factors and such neurohormones as vasopressin and oxytocin. This idea, in some hypothetical form, should be incorporated in the paper.

A couple of weeks later he wrote a follow-up, encapsulating the central idea of the explanation for the transfer phenomenon that he would publish the following year and stick to, convinced that it was entirely consistent with conventional thinking about neural function:

> When a cell assembly fires simultaneously or in rapid succession and, if it is reinforced by a feedback, some material is elaborated . . . which coats or otherwise marks the synapses involved . . . resulting in 'recognition' between the two elements of the synapse.

Indeed, what he was saying was perfectly consistent with what Hebb had written in *The Organization of Behavior*, even though the mechanisms Hebb had suggested were nebulous and speculative only. Ungar hypothesized a precise mechanism with supporting evidence, but his theory had little impact. Scientists, like politicians, sometimes do better by being vague.

By March of 1967 my attention was devoted almost entirely to writing my part of the review chapter with Ungar. We settled on the title "Chemical Correlates of Neural Function," and sought to make the basic point that a chemical dimension overlies and elaborates upon the structural organization of the brain. Ungar was to write about the chemistry of nerve cells and the excitatory process, while I was to write the section on chemical correlates of integrative function, which would bear the brunt of the argument.

A talk by Roger Guillemin about his work on releasing factors at a thyroid conference in Columbia, Missouri on February 18 inspired me to my task. He and others, including Andrew Schally with whom he competed for and ultimately shared the Nobel Prize, knew by then that the releasing factors for the thyroid stimulating hormone and the adrenocortical stimulating hormone were concentrated in the hypothalamus of the brain and were small molecules with the properties of peptides. Years later, after the peptide structure of the releasing factors had been proven, they were shown to modify neural activity of specific circuits and influence certain types of behavior, in addition to stimulating the release of their target hormones from the pituitary gland. At that time, their neuroactive properties were only a supposition, but a supposition central to the point of our chapter and to the way Ungar and I were thinking of the transfer phenomenon.

With the exciting developments on the releasing factors as my centerpiece, I proceeded to construct an argument for the chemical specificity and responsiveness of brain cells. In addition to the neuroendocrine evidence, I relied on the point-to-point specificity of neural connections (implying chemical recognition), the responsiveness of nerve cells to the stimulating action of substances like Levi-Montalcini's Nerve Growth Factor, and their sensitivity to the behavioral influence of specific hormones, to argue that the brain is a chemically coded mosaic responsive to chemically coded information. The transmission of excitation from one cell to another by chemical messengers only scratched the surface of a highly sophisticated chemical system, the full subtlety of which was yet to be revealed.

The transfer experiments were dealt with in my draft in a fairly minor way. While pointing out that "elaborate interpretation of chemical transfer of acquired information must await more data," I noted that the phenomenon could best be viewed "in terms of chemically unique neuronal identities and chemospecific pathways. Transfer of acquired information may simply be transfer of those chemicals which identify specific pathways and synapses by which the information is coded and stored." I mailed my rough draft to Ungar on March 28, 1967, with great relief at finally discharging that burdensome obligation. At about the same time, the National Science Foundation informed me that I had been awarded the Graduate Fellowship for which I had applied, contrary to my expectations. With closure on those high-water marks behind me, I turned my attention to course work, qualifying exams, and a matter of personal moment.

Carol Lee Crumrine, having been graduated from the University of Kansas with a major in chemistry the previous year and with no burning desire to go to work for industry, had been persuaded by her stimulating biochemistry course with Richard Himes to continue her education as a graduate student at Kansas in the fall of 1966. I met her during registration, and was attracted immediately by her beauty, brains, and shy demeanor. As her first year in graduate school wore on, I gradually edged out my competitors for her attention, and by the spring of 1967 was dating her regularly and ready to talk about marriage. She was more hesitant, forcing me to move that issue to the back burner, but resolved to keep it brewing.

New Recruits

Though I was no longer an active participant in the memory transfer field, it was still in the news and it stayed on my mind. While rapidly losing the sympathy of established investigators, the field paradoxically was attracting younger scientists who eventually would do experiments with better controls and more sophisticated designs than the ones first published. Among them were Jim Dyal, an animal psychologist at Texas Christian University in Fort Worth, and his graduate student, Arnold Golub.

Golub was born in Brooklyn, but moved with his family to Los Angeles at the age of 8 because of allergy problems. A number of his relatives died when he was young, so he became interested in science as a means of prolonging life. His tendency to conduct unauthorized experiments in his high school chemistry lab resulted in his banishment to the library for the duration of the course. He therefore entered the University of California at Santa Barbara in 1958 as a biologist without much of a background in chemistry. There he took a course in psychology but found it very boring and cut most of his classes after the first couple of meetings. He discovered, furthermore, that he could make Cs on the exams without even reading the book; so it occurred to him that if he read the book he would probably do very well in the subject. That is how he came to major in psychology.

By his senior year he was working with a physiological psychologist on electrophysiological recording of brain activity. He found the work so interesting that when his professor moved to T.C.U., he followed him there as a graduate student. They had a falling out, however, and Golub ended up working with Jim Dyal, a classically trained animal psychologist who specialized in a learning phenomenon called extinction. An animal that is trained to respond in a particular way, say approach a food cup at the blink of a light by being rewarded with a food pellet for each approach, is said to "acquire" (learn) the behavior. If then no food is provided each time the animal approaches the food cup, in time the animal will stop doing so. The behavior of approaching the food cup is then "extinguished" (unlearned—or in human parlance, "forgotten").

Dyal had read about the transfer experiments and wanted to see if extinction could be transferred. He knew a lot about psychology, but nothing about the brain. Golub by then was doing his doctoral research on hormonally mediated fear responses, a project involving a fair amount of small animal surgery. Since Dyal needed someone who knew how to get the brain out of the cranium of a rat, he asked Golub to team up with him. Golub convinced him that it would not make sense to transfer extinction if, as he expected, they could not transfer acquisition, so they decided to try transfer of acquisition first. But unlike the Jacobson experiment, they had two acquisition groups. One group learned to approach the food cup by the usual method of food reinforcement. A second group was trained in the same way but then extinguished to the point where the rats no longer approached the food cup when the signal light came on; then they were retrained by reinstating the food reinforcement. An untrained control group of donors was also included. To their great surprise, recipients of brain extracts from the retrained group approached the food cup twice as often as the other two groups, an outcome with less than two chances in 100 of occurring at random.

This brought a new twist. For some reason, the brains of rats that unlearned a previously acquired behavior, then were retrained to that behavior, produced something that enhanced the probability that untrained recipients would show the behavior. The same striking results were obtained 3 times out of 4, the one failure involving a new technician. So they felt that there was something to it, but they didn't know what to make of it.

Their first paper was published shortly after a gathering of proponents and some skeptics of the transfer experiments at a scientific meeting in Chicago. Ungar wrote me about the meeting with enthusiasm:

> You should have come to Chicago. We had two evening sessions including all the people who work in the transfer area. There seems to be a considerable change in the wind; the objections turned only around theoretical points such as to decide whether it is really learning we are transferring or some unknown entity. Nobody denied the reality of the phenomenon itself

He added parenthetically that Samson was there, "delegated by F. Schmitt."

Samson remembers that most of the people who came were "believers"— the only significant challenger being Sebastian Grossman. At that time, Samson later would claim, he himself had "some belief in the topic" but not in Ungar. His actions and all his correspondence in 1966 and 1967, however, show nothing but a deep skepticism at best toward the transfer work. In my opinion, Samson was a good enough scientist to sense at an intuitive level that a real phenomenon of some sort—not necessarily memory, but some form of chemically mediated behavioral influence—lurked within the results of the transfer experiments. But his distrust of Ungar was so thorough that he had a hard time separating the phenomenon from its primary proponent.

By that time Samson was responding readily to every communication from me. On June 16, I sent him an outline of the potpourri of experiments I was trying, from which I hoped my thesis research would emerge. These included some experiments on chemospecificity in the retinotectal pathway of chick embryos. There was ongoing methodology for extracting and analyzing gangliosides. And out of my subliminal engineering inclinations, I had designed and was constructing a T-maze to train rats for some learning experiments of the type I had described in my NSF proposal. Lastly I mentioned that I might try a couple of transfer experiments, to clean up some issues left over from the previous summer. My decision to avoid a major investment in the transfer paradigm was still irrevocable (unless I got results too compelling for both of us to ignore). He wrote back immediately with the kind of enthusiasm and encouragement I was coming to expect. "The stuff on gangliosides sounds great to me. I think there is a neurochemical gold mine there."

While I geared up for a major effort on gangliosides, I wanted to take one more shot at the chemospecificity notion underlying the sticky-neuron theory. Byron Wenger, a developmental biologist in the department, agreed to help me test for chemical specificity of cells in the retinotectal connections of the chicken. We chose the retinotectal system because of the high degree of specificity with which nerve cells from the retina, at the back of the eye, connect with nerve cells in the tectum, a region of the brain in birds to which visual information is projected. These connections were known to be topologically rather precise; meaning that a row of 10 cells lined up from top to bottom in the retina would form connections with a row of cells lined up in the same order (except from bottom to top) in the tectum. This implies the existence of a mechanism for recognition between the cells of origin in the retina and the target cells in the tectum. One possible basis for recognition would be the presence of unique marker molecules on each cell, just as a different vegetable specifies location within a vegetable garden. Another possibility would be the patterned arrangement of the same class of molecules according to variable density gradients throughout the cellular field, like the way in which identical black dots produce the image in a news photo merely by virtue of their density and arrangement.

If the mechanism were based on unique chemical identities specific to a given cell or region, we should have been able to induce different antibodies (which react specifically with the chemical structures that induce them) from different regions of the tectum. So the overall experiment consisted of dissecting out specific regions of the tectum from newborn chicks, injecting that tissue into different rabbits which formed antibodies against the injected substances, collecting blood from the rabbits (by bleeding them from small clean razor cuts across the earlobes, which I think hurt me more than the rabbits), and testing the serum from the blood for the presence of antibodies against different regions of the tectum from chickens. This experiment took three months, and in the end the result was essentially negative. No region of the tectum produced antibodies that reacted only with that region. It appeared that the chemical code, if it existed, was more like the arrangement of dots in a news photo than the vegetables in a garden. But the technique in 1967 was still very crude. Barondes, Agranoff, I and others would have greater success later, but from a different angle. Our subsequent work collectively supported the involvement of glycosylated molecules like gangliosides and glycoproteins, or their receptors, like lectin proteins, distributed in a gradient fashion as the basis for chemospecificity. In retrospect, this probably weakened the prospect that individual synaptic connections are uniquely labeled by signpost molecules, but like so many other aspects of the transfer paradigm, that is easier to see now than it was then.

At the end of the summer, Carol with her friend and roommate, Marilyn Hall, flew to New England for a two-week vacation in New Hampshire and New York. Upon her return, Carol's attitude toward me seemed decidedly more receptive, so I was emboldened to raise the subject of marriage once more. This time she agreed, and as she warmed to the idea, our tentative wedding date of somewhere around Christmas got moved up to Thanksgiving; so on November 27, 1967, we were married in her home town of Tulsa, Oklahoma.

Just before Christmas, 1967, Georges Ungar wrote to inform me of an experiment recently reported by Gay and Raphelson, in which rats were given the option of entering a dark box or a lighted box. Being nocturnal creatures, they invariably went first into the dark box, whereupon they were shocked. This turned them quickly into creatures that tolerated the light box quite nicely. Ungar had repeated this "dark-avoidance" training, and found that he could transfer what he called fear of the dark quite well. Excited by my own experiments and preoccupied with the events surrounding my marriage, I paid little notice to this shift in his research. That shift, however, profoundly affected the transfer field. Encouraged by the robustness of these results, a number of researchers turned to the dark-avoidance task as a means of pursuing the transfer phenomenon. This included Ed Bennett, who remained skeptical; Bill Byrne, who believed there was something to it; and Arnold Golub, who had expected to find little but seemed to be finding a great deal.

My own experiments were on a different track. Though fascinated still by the transfer work, I had succeeded in cutting out a less controversial path that I thought had about as much chance of success in the long run as the transfer approach. Ungar thought my experiments too mundane, and Samson considered them a little exotic. I was somewhere in between, recognizing with increasing clarity that the search for memory molecules, as such, was futile and that even the discovery of unique and specific molecular correlates of memory storage was a long shot with the methodologies in use at the time. But no science course I had ever taken had taught me not to take chances. And I didn't see anybody doing anything better.

1968

For me, the nostalgia of the Sixties includes nearly nothing from 1968—a year rocked by political assassinations (Robert Kennedy and Martin Luther King), urban violence, increasing casualties and setbacks in Vietnam, and depressing divisions across generations, races, and cultures. Despite the distractions, I tried to keep to my personal agenda of finishing my doctoral research as soon as I could. Having passed my oral qualifying exam, it was possible for me to finish in a little more than a year, so the summer of 1968 was beginning to look like my last as a graduate student. Samson was aware of this, and it posed a problem for him. Frank Schmitt had asked him to return to NRP for the summer, and Samson badly wanted to go. But that would mean leaving me without a doctoral advisor again, for several months at a critical time. Samson's solution was to ask if I could come to Boston to NRP with him. He cleared it through Schmitt and posed the possibility to me in January. I was not enthusiastic, as already by then I could see how much I still had to get done before starting to write my dissertation. Furthermore, I had only been married a couple of months and knew that Richard Himes would not think much of having Carol, who by now was doing her own dissertation work in his lab, take off with me on what he probably viewed as a summer-long junket to Boston. Finally, exactly what I would do at NRP was not at all clear. On the other hand, I had never been east of the Mississippi River to speak of, and Boston had an undeniable, exotic appeal. Carol and I thought about it for some time, and came up with a compro-mise to propose to our respective advisors: I would go for two months instead of three, and Carol would join me for one of those two. This was acceptable to Samson, and grudgingly accepted by Himes.

Having resolved the problem engendered by my traveling advisor, I felt greater pressure than ever to get my work done as the Spring arrived. The pace of my research picked up accordingly, and the season passed in a blur of long hours in the lab against a depressing backdrop of social and political turmoil, culminating in the assassination of Martin Luther King on April 4, 1968.

Bill Byrne had returned to Duke from Berkeley, convinced of the importance and validity of the transfer experiments, though unsuccessful in persuading his colleagues in California on that score. It had come time to choose a new

Chairman of Biochemistry at Duke, but Phil Handler made it clear that Byrne would not be the one selected. Not wishing to stay there under another chair, Byrne sought positions elsewhere, and was offered the chair of Biochemistry at the University of Tennessee Health Science Center in Memphis. He was set to finalize the move the week that King was assassinated. After long and hard thought, he decided to proceed. Staying at Duke was not a viable option, and he didn't see why he couldn't do good science in Memphis, so he went ahead with the move.

I mitigated the depression which the King assassination and subsequent violence in cities across the country had aggravated by deciding to fly to the Federation Meetings in Atlantic City for the first time. Atlantic City I found a bit tawdry, but I wasn't there for the scenery. That evening a symposium was held on the current status of the memory transfer field. Bill Byrne chaired the session, which featured Ungar, Bennett, Rosenblatt, and Fjerdingstadt. Ungar and Fjerdingstadt presented data that I characterized as "near-unequivocal" support for transfer of learned behavior. Rosenblatt's results "as usual, were very complicated, but tended to support positive transfer." Bennett, the skeptic, presented inconsistent data. Everyone's performance was true to form.

After the formal session, we adjourned to a pub around the corner, where I got to know better the men who already were legends to me—Bennett and Rosenzweig—and the newer additions to the fraternity of molecular memory research: Eijnar Fjerdingstadt, Frank Rosenblatt, Bill Herblin, and Bill Byrne. It was Byrne who impressed me the most that first night that I met them all. His lack of pretentiousness, his earnestness, and his sincerity convinced me totally that he just wanted to get at the truth. Rosenblatt struck me in a similar way. Ungar, Bennett, and Fjerdingstadt, of course, were sincere and honest scientists, but I had the feeling that they already felt they knew where the truth lay. It wasn't till 2:00 in the morning that Ungar, Herblin, and I headed back to our hotels, reviewing with enthusiasm the evening's discussion of molecules and memory as we walked down the fabled but deserted boardwalk.

It was the last time I talked to Ungar at that meeting, though the next day I caught a glimpse of him at a distance, walking by himself, looking oddly alone. It troubled me to think that he was staking so much on a dubious phenomenon accepted by no one of stature and reputation comparable to himself.

My summer in Boston turned out to be surprisingly productive. Assigned to a bleak corner office in the huge Brandegee Estate mansion that at the time was the home of the NRP, I was given the task of looking over material collected in a Work Session a number of months earlier on "Aspects of the Brain Cell Environment." It had dealt with the chemical structure and physiological function of the surface of cells in the brain and the space between brain cells. Since gangliosides were thought to be localized at the cell surface, with their carbohydrate portions sticking out into the spaces between adjacent cells, and since they and their possible physiological functions were in fact a subject of discussion at the meeting, the work session was of inherent interest to me. It certainly would provide useful background material for my dissertation. Samson must have seen this in getting the topic assigned to me. As the work progressed, we struggled with the question of exactly what constitutes the "environment" of brain cells. The focus of this work session was really on the way in which the chemistry of the cell surface influences the bioelectrical properties of the inter-cellular fluid right next to the cell; and, alternatively, how that milieu influences the excitability, hence function, of the cell. It was the *micro*environment of the cell surface and the space immediately adjacent to it that was the real focus of concern. I therefore suggested that the work session be entitled "Brain Cell Microenvironment." Samson and the rest of the NRP staff liked the idea, so that became the permanent title, and after the publication of that issue of the *NRP Bulletin*, the term 'microenvironment' came into general usage in neuroscience.

By this time Carol had arrived. She was appalled at our housing, but we couldn't afford anything else I could find, so we decided to endure the month in our one room apartment with hotplate and half-refrigerator off Huntington Ave. Lacking a television, we rented one to watch the Republican convention, on the eve of which I told anyone who would listen that Nixon played well in Winona but could never beat Rockefeller in the heavily-populated urban centers of the Northeast. Having lived in one for a month, I considered myself an expert on urban politics. Notwithstanding my acumen, Nixon swept to victory on the first ballot in the early morning hours of August 8.

Meanwhile, a manuscript on the Brain Cell Microenvironment Work Session was taking shape, to the apparent amazement of the staff. As that issue of the *NRP Bulletin* continued to gestate, the Hong Kong flu swept the nation. Carol and I caught it, and nearly died unknown in our one-room flat far from home. It was good fortune that our fevers peaked at different times, so that the one who was merely sick could minister to the one who was truly ill. Czecho-slovakia was invaded by troops of the Warsaw Pact. In Chicago, the Democratic National Convention opened in the atmosphere of an armed camp. I finished a draft of the Brain Cell Microenvironment issue of the *NRP Bulletin*. Heat descended on Boston with a vengeance, and Carol and I finally fled to the mountains of New Hampshire where I composed an essay on the implications of the recent research on the brain cell microenvironment. I sent it to NRP, which eventually published an essay written with greater flourish but less clarity by Schmitt instead.

Our short season in New England thus completed, Carol and I drove back to Kansas the long way around, through Houston and San Antonio. Ungar was having a great deal of success with his new dark-avoidance task, the one to which I had paid little attention when he first mentioned it in a letter the previous December. In March, he had published results with this new behavioral task, and the evidence for transfer of dark-avoidance into mice not previously trained to avoid the dark was so dramatic that statistical methods for analyzing the differences were not even necessary. Finally he had found a behavior that could be transferred with enough reliability and consistency that he felt he could go to the next step—the purification and identification of the substance in the brain extracts responsible for the altered behavior of recipients.

For the next two years, Ungar and a small group of hard-working assistants labored at extracting and purifying peptide extracts from thousands of rat brains, in the quest to isolate and chemically characterize a specific peptide that by then he felt was responsible for predisposing mice to avoid the dark with no previous experience that would have impelled such behavior. On a lower floor at the Baylor College of Medicine, exactly the same effort was underway to isolate releasing factors from thousands of pig brains for the anterior pituitary proteins whose release by then was known to be triggered by peptides from the hypothalamus. While the latter effort would ultimately yield a Nobel Prize for Roger Guillemin, Ungar's fate would be less fortunate.

Though vaguely aware of all these activities, I was oblivious to their details. My mind was focused on completing my doctoral research and getting the dissertation written in time to be graduated with a PhD on June 1, 1969. After that, I concentrated on a year of post-doctoral study at the Parsons State Hospital and Training Center, where I helped set up a neurochemistry lab while continuing to work on ganglioside metabolism and brain stimulation in Samson's lab in Lawrence. Commuting between the two locations enabled Carol and me to continue to live together with frequent but brief absences as she completed the work for her doctoral degree.

By the early Spring of 1970, with Carol's doctoral research near completion, we were anxious to move on. In response to one of many inquiries concerning faculty openings at colleges around the country, the chairman of the Department of Biology and Pharmacology at the College of Pharmaceutical Sciences, Columbia University, invited me to New York for an interview. Within days of my return to Lawrence, the offer of a position as Assistant Professor of Biology came through. A brief exposure to the exotic taste of New York during our summer in the Northeast in 1968 left Carol and me anxious to leave mid-America behind in favor of life in the big city. So without waiting for any other offers, I accepted immediately.

Climax

Dead on Arrival

On his twenty-ninth birthday, Dominic Desiderio awoke to find his picture on the front page of the *New York Times*. It was January 11, 1970, shortly after he and a team of researchers headed by Roger Guillemin at the Baylor College of Medicine in Houston had announced their discovery of the chemical structure of thyrotropin releasing factor (TRF). The compound is a peptide manufactured in the brain for transport to the anterior pituitary gland, where it stimulates the release of the hormone that controls the function of the thyroid gland. The chemical characterization of the first of the releasing hormones was a finding of major importance, because it opened the door to a whole new class of informational molecules in the brain.

The vast majority of progress in science doesn't come in the course of a mad dash for fame and fortune; but pinning down the chemical structure of TRF was the culmination of perhaps the greatest scientific race since the discovery of the structure of DNA. As it became clear by the mid-Sixties that most of the releasing factors were probably peptides, two labs emerged at the forefront of the effort to chemically identify them. One was headed by Andrew Schally at Tulane, the other by Roger Guillemin at Baylor. Desiderio had been recruited to Guillemin's team because of his expertise in the sophisticated science of mass spectrometry—an instrumental method for deciphering the chemical structure of molecules. Schally and Guillemin had narrowed their focus to TRF—it was apparently a small molecule (only three amino acids, as it turned out), and seemed most likely, therefore, to be the easiest to identify first. But an unusual structural feature delayed the solution far beyond what either Guillemin or Schally had expected. Both labs worked at a fever pitch, publishing every least, little advance to maintain their position in the race. It was not unusual for Desiderio and his colleagues to go over their latest results in Guillemin's living room on a Thursday night, write up a brief report and telex it to Paris on Friday morning, and have the paper appear in the *Comptes Rendus de L'Academie de Sciences*, the journal of the French Academy of Sciences to which Guillemin had access, within the next week.

Dominic Desiderio was apparently born to be a scientist. By the eighth grade he had taught himself the basics of chemistry from his home encyclopedia and had decided that he wanted to get a PhD. Raised in the Pittsburgh area, he excelled academically in high school, winning a prestigious Westinghouse Science Talent Search Scholarship to the University of Pittsburgh in 1958. Graduating three years later with a bachelors degree in math and chemistry, he moved on to M.I.T. where he completed his doctoral work in analytical chemistry in 1965, working with Klaus Biemann who was just beginning to probe the molecules of life with mass spectrometry. After a couple of years at American Cyanamid, he was recruited by Baylor to join the faculty of its Institute of Lipid Research. There he worked on computerizing mass spectrometry, on technical innovations and instrumentation, and on lipids such as prostaglandins; but he avoided peptides like the plague because of their chemical fragility and complexity. In time, however, his friend, Roger Burgus, finally persuaded him to join the quest of the Guillemin team for the structure of TRF. His expertise was a decisive factor in Guillemin's victory, which they knew they had when they put the final pieces together about 9 o'clock one Thursday night in late 1969. It was a grand moment of supreme satisfaction.

With the glorious feel of success still fresh on his mind, his threshold was lowered when he attended a seminar given by Georges Ungar a few weeks later on another peptide of considerable potential importance. Ungar by then had isolated a tiny amount of scotophobin, the peptide he claimed could induce dark avoidance in mice. But he had reached the limits of his technical expertise and didn't quite know how to get to the next stage—namely, working out the chemical structure (amino acid sequence) of the peptide. Desiderio was struck by what little amount of scotophobin Ungar had to work with, but offered to help Ungar as best he could.

Their initial attempts to obtain spectral data from tiny samples of the precious material were inconclusive. Peptides are notoriously susceptible to breaking apart, either in the process of extraction or the analytical phase, and Desiderio could tell that the scotophobin they were working with was broken into several pieces. With computer analysis, he was able to piece together structural permutations for the different fragments into a few likely structures for the entire intact molecule. That, in combination with Ungar's previous more conventional biochemical methods enabled them to narrow the likely amino acid sequence to just two or three possibilities. Wolfgang Parr, a chemist at the University of Houston, synthesized the likely sequences, and one of them had the same dose-response influence on dark avoidance, and the same chromatographic behavior, as the naturally extracted compound. These two pieces of evidence were taken by Ungar, Desiderio, and Parr to constitute proof of the chemical structure of scotophobin. They made the fateful decision to submit the paper to *Nature* for publication, and Desiderio prepared for his second scientific triumph in as many years.

This time, though, a quirk of publishing fate intervened. Since much of the evidence for the chemical identity of scotophobin rested on spectrometric data,

Nature selected as a reviewer an expert in analytical spectrometry at NIH by the name of Walter Stewart

Walter W. Stewart was not a conventional scientist nor an ordinary person, so it is not surprising that his scientific career had traced an unusual path. Born in New York City in 1945, he entered Harvard at the age of 18, was elected to Phi Beta Kappa in his senior year, and graduated with a BA Summa Cum Laude. He spent 1967-68 as a Graduate Fellow at the Rockefeller University, departing without a degree to a "Commissioned Office" at NIH for two years. From 1972 to 1974, he was a Junior Fellow in the Society of Fellows at Harvard University, and a "Guest Scientist" at NIH. In 1974 he became a Staff Fellow, and in 1979 became a Research Physicist at NIH without benefit of a graduate degree from either of the prestigious institutions he had attended.

When he reviewed Ungar's paper on scotophobin, he had only two refereed publications, but one of them was on the "Isolation and proof of structure of wildfire toxin" in *Nature*. The chemical structure of wildfire toxin is unusual for a natural compound, and his difficult proof of its structure demonstrated ample expertise to judge the claim of the isolation and proof of structure of scotophobin. The manuscript from Ungar, Desiderio, and Parr arrived at *Nature's* editorial office on 8 February 1971. There ensued over the following year an extraordinary correspondence between Stewart and the authors in Houston, consisting in essence of Stewart asking for more data and details, with Ungar responding in part but never fully to Stewart's exacting demands for detailed information. By the Spring of the following year, it appeared that agreement between authors and referee was impossible, so *Nature* made the highly unusual decision to publish the four page paper submitted by Ungar, Desiderio, and Parr back to back with a seven page critique by Stewart, and a one page rebuttal by Ungar.

Stewart's criticism focused most heavily on the alleged purity of the scotophobin fraction, and the mass spectrometric evidence for its chemical structure. Stewart's assertion that the material was too impure for reliable quantitative analysis depended on a selective interpretation of the data, and appears to have been overstated. His critique of the spectrometric evidence was largely hypothetical and stripped of the context of other evidence obtained by Ungar and his colleagues from conventional biochemical procedures. Nonetheless, his expertise for evaluating this evidence was unquestioned, and he did make some telling points. "It was good," Desiderio would later admit, "because he honed right in on the things that should have been honed in on."

This generous respect of one spectrometrist for another was not shared by the authors collectively, however. Their rebuttal began, "It would be impossible within the limits of five days and 1,500 words granted to us, to answer in detail the criticisms for which Mr. Stewart was given over a year and apparently unlimited space." They did try, however, in the one page allotted for their response, to repair the damage dealt by Stewart's overwhelming quantitative, statistical, and hypothetical arguments. Given the tiny amounts of the natural scotophobin they had had to work with, though, the ambiguity of the analytical chemical data was inevitable, and the contrived though logically valid argument by Stewart could not effectively be rebutted.

To Ungar, this was all beside the point. In his mind, he had accomplished an incredibly difficult isolation of a natural product, established a 'probable' chemical structure with the best conventional techniques available at the time, and obtained analytical data consistent with that structure. That some weaknesses could be found in any given link in the exhausting chain of research that had led to that point, was no surprise at all. The point remained that the world had been given a specific chemical compound—a precisely defined sequence of 15 amino acids—whose significance could now be fully explored.

That point was lost in the neuroscientific community at large, however. Worried at the possibility that a major discovery had eluded them, and confused by a torrent of conflicting publications, most neuroscientists were relieved to embrace the seemingly erudite and objective refutation by Stewart of Ungar's claim to have discovered a unique and specific molecular representation of experience in the brain. The paper by Ungar, Desiderio, and Parr on the chemical structure of scotophobin was dead on arrival. The day was just around the corner when small peptides on the order of scotophobin would be recognized to play a central role in the processing of neural information, but the claim would come from a different, more credible direction, and Ungar would play no part in it.

Later in his career, Stewart would make some significant scientific contributions, including the discovery of Lucifer dyes—extremely valuable compounds for tracing neuronal pathways. He and his colleague, Ned Feder, however, would gain greater fame as crusaders against scientific fraud, and would institutionalize themselves as irritating whistle blowers at NIH. The gadfly image for Stewart carried beyond the lab. He would get in trouble, for instance, with the municipal authorities of his suburban Washington neighborhood for letting his lawn turn to a field of weeds. This all lay in the future, however, and none of it necessarily draws into question his qualifications for refereeing the paper on scotophobin. But to the extent that Ungar's history of flirtation with controversial ideas is relevant to the fate of scotophobin, it seems only fair to point out the impulse toward iconoclasm displayed by the person who ultimately turned out to be one of Ungar's final and fatal critics.

Let the Test Tube Tell the Tale

In March of 1971, with the protracted frustration and ultimate stillbirth of the scotophobin paper in *Nature* still to come, Georges Ungar was a celebrity in the popular press, and briefly among a sizable segment of the scientific community. The American Society of Neurochemistry held its second annual meeting that month in Hershey, Pennsylvania; and Georges, with his news that the scotophobin molecule had been sequenced, was one of the primary topics of conversation. He chaired a symposium on the transfer work, in which he invited me to participate as a discussant. Though the neurochemists in attendance were polite, many were clearly skeptical, and I had the feeling that the pitch that Ungar and I were making for proceeding to the next research step (analyzing in greater detail the source and behavioral specificity of the transfer factors, and seeking to identify other peptides with behavior-altering properties) was not falling on particularly sympathetic ears.

Even less sympathetic and distinctly less polite was the audience at another symposium, featuring the work of another controversial figure at the time, Samuel Bogoch. Bogoch was one of the first, along with Barondes, to seriously consider the possibility of an informational role for glycoproteins in the nervous system. This was certainly consistent with my notion of the importance of the extracellular domain of nerve cells, where glycoproteins and gangliosides had to play a critical role, so I was sympathetic to his ideas. Unfortunately, Bogoch—a far more eccentric personality than Ungar—was a zealous proselytizer who didn't publish his work in refereed journals. His claims of biochemical changes associated with operant conditioning in pigeons were greater in magnitude than most workers found credible; and there were technical details of his research that nearly all neurochemists, including me, found highly suspect. The symposium erupted into an acrimonious debate between Bogoch and his critics. I suggested that the best way to clarify questionable research was for those of us who thought the basic ideas might have merit, to repeat Bogoch's experiments and let the test tubes from many labs tell the tale. This, I would eventually try to do, and probably not to my benefit.

The next year, the American Society of Neurochemistry met in Seattle. Doing research at the College of Pharmaceutical Sciences in New York had proven difficult for me, as the facilities were antiquated and the teaching load was heavy. But I had managed to do some basic work on the response of rat brain sialidase, an enzyme that regulates the composition of glycoproteins and glycolipids, to behavioral stimulation, and this was the work that I presented at the meeting. I did keep an eye on Ungar's progress, however, and in the eyes of my mentor and colleagues, was still associated primarily with research on the neurochemical basis of behavior. Over drinks with Fred Samson and Stan Twoomey, another of his students, I found it curious when they both asserted that my type of research was the "really important" work in neurochemistry. If this is so, I asked, why aren't more people doing it? The bottom line, we decide-ed, was that it was simply too risky in terms of yielding both publishable and interpretable data. I should put the emphasis on "interpretable," perhaps, because behavioral neurochemistry research was certainly being published then— it just wasn't clear what kind of sense it made. Even if I *had* found significant differences in sialidase activity between trained and untrained rats, what would it have meant?

One idea I definitely *did* believe in, however, was that the disparate and sketchy information on enzymes that control the metabolism of the glyco-conjugates should be brought together in a search for broader generalizations. Thus I attended an evening "seminar" on glycolipid enzymes with anticipation, thinking that just such an integration might be discussed. The group in attendance was a good representation of leaders in the field of glycolipid metabolism, including Abraham Rosenberg, Julian Kanfer, Kunihiko Suzuki, Joseph O'Brien, Joel Dain, and the chairman of the session, Norman Radin (whose role in the confrontation between Samson and Ungar fourteen years previously was des-cribed earlier). By the end of the evening, though, not one person had even attempted any synthesis. Instead, each panelist got up and talked about their own specific experiments without any attempt to draw connections, contrasts, or parallels with any other work. Already frustrated by the lack of response to my behavioral paper earlier in the week, this concentration on the trees instead of the forest was more than I could take. I gained the floor and suggested that the collective wisdom and expertise in the room at that point was too great to waste without attempting to at least come to some generalizations about glycolipid enzymes as a family. No one offered any. I then sketched out of table, with columns for different properties such as pH requirements, subcellular locali-zation, and so forth; then invited everyone to fill in the information for his or her pet enzyme. No one was interested in doing so. I asked if anyone thought such a tabulation would be useful, and the group indicated "not particularly."

"Then why do these experiments at all?" I asked.

Norman Radin shrugged and said, "Each man to his own thing."

Grinding and Binding Gets Going

By 1972, American battlefield deaths in Vietnam had dwindled to a few hundred a year, down from a peak of over 14,000 in 1968. But the war was taking a dreadful toll on American personnel in another way. The government estimated that up to 30% of American soldiers were using heroin by the time they returned from their tour in Vietnam. In the wake of these alarming statistics, and in view of the spreading blight of heroin addiction domestically, the Nixon administration decided to wage a "war on heroin" by committing a quarter of a billion dollars to methadone maintenance programs and basic research on drug addiction. Among the beneficiaries of this windfall were Avram Goldstein at Stanford, Solomon Snyder at Johns Hopkins, and Hans Kosterlitz at the Marishal College in Aberdeen, Scotland. These three pharmacologists and their colleagues, along with a handful of others, were about to play a decisive role in the way that brain peptides were perceived and studied by the vast majority of neuroscientists.

From the published record, it would appear that Ungar's pioneering experiments on transfer of morphine tolerance, followed by his more dramatic work on the biochemical transfer of behavioral tendencies as epitomized in the discovery of scotophobin, had remarkably little influence on the research that convinced everyone of the reality and importance of neuroactive peptides. The line of investigation that proved convincing burst upon the scene of neuroscience explosively in 1973, with the unequivocal demonstration of opiate receptors in the brain, followed by the discovery in 1975 of naturally occurring compounds in the brain with opiate-like properties. These endogenous opiates turned out to be peptides. But in very few of the reports of those discoveries was the work of Ungar cited; and after 1980 it became uncommon in reviews of the literature on neuroactive peptides for Ungar to be mentioned at all.

Were the discoverers of endogenous opiates and their receptors unaware of Ungar's work? Hardly. In the decade between 1965 and 1975, only the most reclusive of scientists could have escaped the publicity attendant to the behavioral transfer paradigm. Ungar's long career had made him a well-known pharmacologist, many of whose practitioners knew him personally. While some may have been unaware of the history of his work on morphine tolerance that led him to test for the transfer of sound habituation, those who were working on the mechanisms of analgesia and drug addiction were certainly aware of it. To what extent was their research influenced by Ungar's work? The answer to that is a little more complicated. Avram Goldstein is an interesting case in point.

Goldstein has to be considered a central figure in any discussion of the discovery of opiate-like peptides. The son of a prominent rabbi, he was born and raised in New York City. Walter Sassaman, a teacher at the private uptown school that Goldstein attended, first kindled his interest in science. An honors student in biochemistry at Harvard College, he was graduated from Harvard Medical School in 1943, having spent much of his time cutting classes to work on research. After the war, he became an Assistant Professor at Harvard, from which he was hired to be chairman of Pharmacology at Stanford in 1955. One of his earliest research interests had been the phenomenon of drug tolerance, but at Stanford his work had moved more in the direction of molecular biology, where exciting discoveries were unfolding with rapidity at the time. He wrote a definitive textbook on *Principles of Drug Action*, and played a role in moving pharmacology in a more molecular direction. By the late 1960s, he had tired of administration and was looking for a new research focus. With the ample influx of more potent heroin from Southeast Asia fueling a drug epidemic in this country, he decided to return to his earlier interests in drug tolerance and its relationship to addiction. A federal grant of $400,000 from the "war on heroin" program helped him establish his own Addiction Research Foundation. Interested in both clinical treatment and basic research, he helped set up methadone programs, and turned his energy toward laboratory investigations on the mechanisms of opiate drug action (hence addiction as well). This brought him unavoidably to an acquaintance with Georges Ungar.

While Goldstein's preferred approach was much more mechanistic and biochemical, Ungar's results with behavioral assays were too dramatic to ignore. Thus, Goldstein attempted to replicate Ungar's claim of the transfer of morphine tolerance by brain extracts. He apparently made a substantial effort, but had no success. In a note to *Nature* in 1971, he reported his inability to replicate Ungar's results, but was gracious in acknowledging that he could be wrong. In the meantime, he focused his attention on a more cellular, biochemical approach—namely, the detection of naturally occurring molecular receptors for the opiates. About this time he devised the strategy that, with slight but critical variations, would ultimately prove successful in the hands of others. In short, he measured the amount of radioactive opiate drugs that would "stick" to homogenates of brain tissue. The key to his method was a chemical strategy for distinguishing between the amount of drug bound specifically to an opiate receptor as opposed to the amount bound nonspecifically to other cellular components. Using this strategy, he found that a disappointingly low level of binding—only about two percent above background—occurred in brain tissue. Nonetheless, he published the technique in the *Proceedings of the National Academy of Sciences* in 1971, offering it as suggestive evidence for the presence of opiate receptors.

Again Goldstein was diverted by Ungar's claims, this time of the discovery of a fear-inducing peptide, scotophobin. Once more, he attempted to replicate Ungar's work, and again he failed. In frustration, his skepticism turned to cynicism. With a growing awareness of Ungar's controversial past, he began to suspect fraud, and determined to unmask it by visiting Ungar's lab and watching his entire procedure from start to finish. Ungar welcomed his skeptical visitor for a week of observation. Goldstein had total access to Ungar's labs, his technicians, and his results. He watched the animals being trained and tested, observed the biochemical extractions, and followed the analysis of data. He could detect no evidence of fraud. When he returned to California, however, he still could not repeat the results that Ungar obtained, albeit with some modifications of the procedure. Again he dispatched a note to *Nature*, which appeared in 1973. This time his tone was perceptibly more impatient and hostile. He had had it with Ungar and the whole idea of biochemical transfer of anything from one animal to another. Yet by then he had certainly begun to think of the possibility of endogenous opiates, having started a notebook on the subject in 1972. And Ungar had asserted over eight years earlier that the active principle responsible for transfer of morphine tolerance is a peptide.

In the course of his considerable efforts to repeat Ungar's experiments, Goldstein also undertook a laborious attempt to isolate the opiate receptor. These time consuming projects took him away from the drug binding strategy that he had instigated. Meanwhile, others made minor adaptations to his procedure, and began to make headway toward demonstrating the presence of opiate receptors much more definitively than he had done in 1971. The first appears to have been Lars Terenius, a young professor at the University of Uppsala in Sweden. Terenius was well acquainted with the binding methodology, having made important contributions to the demonstration of intracellular receptors for the female hormone estrogen. One of his tricks was to use a ligand (the molecule whose binding to tissue is being monitored) with much higher specific radioactivity—that is, with a much higher proportion of the molecules that carry the radioactive tag—than Goldstein had employed. In early November of 1972, Terenius submitted a brief report of his evidence for specific opiate-binding activity to a Scandinavian journal, where it languished *in press* for six months while two labs in the United States closed in on the same conclusion.

Solomon Snyder at the Johns Hopkins Medical School in Baltimore also received $400,000 of "war on heroin" money, by proposing to pursue the search for opiate receptors as suggested by Goldstein, whom he had heard explain the methodology at a conference in the summer of 1971. Candace Pert, a new, enthusiastic, imaginative graduate student in his lab, assumed the project. Pert and Snyder had the good fortune to work next door to Pedro Cuatrecasas, who had pioneered a rapid filtration technique for doing binding studies. Samples of tissue homogenate were placed in a large number of small wells, each with a glass-fiber filter on the bottom. The radioactive ligand was added to the homogenate, then the mixture was sucked through the filter quickly after a precise, brief incubation period. The method was simple, fast, and enabled the running of a large number of samples simultaneously. Throughout the summer of 1972, Pert made little progress. Then, in mid-September, she tried a batch of highly radioactive naloxone, a powerful opiate antagonist that worked. The amount of specific binding was dramatically higher than she had ever seen before, or than Goldstein had ever seen in any of his experiments. Other drugs were quickly tested, and their binding capacity was found to correlate with their analgesic potency. Evidence for binding specific for an opiate receptor at last was robust and unambiguous. Their paper announcing these results was submitted to *Science* in December, about a month later than Terenius had submitted his.

Eric Simon, a professor at the New York University Medical School was hard at work on the opiate receptor problem at the same time as Terenius and Pert, apparently unaware of their efforts until the Pert-Snyder paper appeared in March, 1973. As far back as 1965, Simon had conceived the idea to test for opiate receptors using radioactive nalorphine, but the effort had not succeeded and he had turned his attention to research on bacteria. Then Goldstein's 1971 article on how an opiate receptor could be detected in principle had rekindled Simon's interest, and by the fall of 1972 he and his colleague, Jacob Hiller, were seeing what they considered good evidence for the specific binding of etorphine, a narcotic thousands of times stronger than morphine. Though their work had not been published when the paper by Pert and Snyder appeared in *Science* in March, a presentation of their work had been scheduled for the Federation meetings in April. Thus Snyder and Simon more or less simultaneously, and independently, confirmed the existence of opiate receptors in the brain, as suggested by the earlier but somewhat weaker evidence of Terenius, and as hinted at by the groundbreaking experiments of Goldstein.

To Hans Kosterlitz and John Hughes in Aberdeen, the news of the discovery of opiate receptors came as no great surprise, for they were already at work on the next step—identification of the naturally-occurring endogenous substance for which the receptors presumably existed. Kosterlitz had more in common with Ungar than any of the other principles in narcotics research at the time. European by birth and training, he came to research as Ungar had from the study of medicine, with a broad background in old-fashioned physiology and pharmacology, which included the use of bioassays. Most notably, he was an expert on the use of the guinea pig ileum, the same preparation that Ungar had used to help discover the first antihistamines. Kosterlitz's discovery that narcotics would inhibit contraction of the guinea pig ileum, and that the effect could be blocked by naloxone, provided the basis for the most widely used test for screening for the potency of opiate drugs and their antagonists.

Hughes began initially by screening a number of brain extracts from guinea pig brains using muscle contraction of the vas deferens as a bioassay. After he detected a small positive response, he switched to the guinea pig ileum and began to test large-scale extracts of pig brains obtained from the local abattoir. Though a veteran of research on angiotensin, a peptide that affects constriction of blood vessels, Hughes did not think it likely at first that the endogenous opiate-like substance from the brain would be a peptide. It was Terenius, who had joined the lab in Aberdeen for a brief period at the invitation of Kosterlitz, who believed on the basis of experiments that he had been conducting along a similar vein that the substance more resembled a peptide than it did anything like the structure of the known opiate drugs. He persuaded Hughes to incubate his fractions with peptidase, an enzyme that breaks down peptides specifically. When the fractions thus treated lost their bioassay activity, everyone in the lab became convinced that they were on the trail of a peptide, and probably a pretty small one. It would take another year and a half of work, culminating in the use of the same mass spectrometric technique that Desiderio had painstakingly applied in the face of Stewart's stinging critique to the solution of the chemical structure of scotophobin, before an endogenous peptide with opiate behavior would be revealed. In the mad dash toward that discovery, Ungar's scotophobin was all but ignored. Neuroscientists found strips of smooth muscle twitching in a salt water bath easier to interpret than the erratic behavior of mice that behaved, on occasion, as though they were afraid of the dark.

"This Controversy Won't Be Settled by This Generation"

In the summer of 1972, as Candace Pert struggled to establish the existence of endogenous opiate receptors in the brain, David Malin moved from Ann Arbor to Houston, where for a brief time with Georges Ungar he would fan a few flames out of the dying embers of the scotophobin research.

David Malin was born on June 9, 1944, in Washington, D.C. A gifted student, he did well enough at Walter Johnson High School in Bethesda to gain entrance into Harvard in 1962, where he gravitated toward the science of behavior. One of his professors, fascinated by the growing preponderance of hippies in Harvard Square, sent Malin out into the streets of Cambridge to interview the newly committed legions to an alternative life style. It was this experience that attracted the attention of professors at the University of Michigan, where Malin had applied to graduate school because of the reputation of the Mental Health Research Institute there.

His first research at Michigan, therefore, was in the psychodynamics of personality—a continuation of his exploration of the drug culture begun at Harvard. But his ultimate interest was in the mechanism of information storage and retrieval in the brain, and psychodynamics was too subjective and elusive a tool for getting at something that he felt had to be more objective and specific. "We all seem to have a preferred level of ambiguity in our research," he observed; and the level of ambiguity in the field of psychodynamics was too great for comfort to him. So he transferred to the biopsychology program, taking up a project on how brain lesions affect conditioned reflexes. But if you were in Ann Arbor in 1967 and wanted to study learning and memory, your only compelling choice was between James V. McConnell and Bernard Agranoff. In a decision that was fateful to say the least, Malin chose McConnell.

Soon after his transfer into biopsychology, Malin met Arnold Golub, who by then was a postdoc with Jim McConnell.

"Oh," said Malin, "Are you the 'Golub' of 'Dyal and Golub?' " referring to Golub's publication of his transfer work with Jim Dyal at TCU.

Golub, surprised and impressed that a stranger recognized his name, was happy to bring Malin into the McConnell fold. Together, Golub and Malin began the task of trying to transfer in rats both negatively and positively reinforced behavior—learning that was induced in one group by punishing wrong choices, and in the other group by rewarding right choices. Malin entered into this effort frankly expecting the experiment to fail, or at least to give results with a trivial explanation. Instead, the results were robust and best explained only by the assumption that information specific to the method of training had been transferred chemically from donor to recipient.

"If I had been lucky enough to get negative results, my whole career would not have been ruined," he would later lament, with only slight exaggeration.

Malin and Golub made a good team, and did probably the best research to come out of McConnell's lab between 1967 and 1970. Both were far less flamboyant than McConnell, far less prone to gimmickry and hyperbole; and they were independent researchers in their own right. McConnell appreciated that fact, and encouraged them to publish without his name, but they felt that it wouldn't be right.

"In his own way, McConnell was a rather careful researcher," Malin explained. "I know that in some quarters, he doesn't have that reputation. But he was careful about experimental design, he was careful about controls, he was careful about accommodating experimental design to species differences because of his background in experimental psychology."

So out of respect, McConnell's name was included on their papers, even though— like all the transfer work associated with McConnell—it was hardly a benefit to them in the long run.

About midway through Malin's work on his doctorate in 1970, Ungar came for a visit to McConnell's lab. Conforming initially to McConnell's bias that the transfer material was nucleic acid, Malin and Golub had used the cold phenol extraction method for RNA to obtain transfer material from donor rats. But by the time of Ungar's visit, Malin had bought into the latter's argument that the active principle was carried by peptides contaminating the RNA fraction. Malin had also become convinced that Ungar was the most dedicated researcher in the transfer field, and the one most likely to achieve ultimate success. Upon hearing all this from Malin, Ungar pulled out a vial of the material that ultimately would be named scotophobin, and offered it to Malin for testing. Malin apparently stared at it for a long time, relishing his good fortune, because Ungar admonished him not to "look at it like it's the holy grail."

In the early flurry of excitement over the transfer experiments, many scientists and would-be scientists took up the quest around the world. Prominent labs, already with well-established reputations, attempted to replicate the transfer phenomenon. According to both Malin and Golub, this was true of labs at Michigan other than McConnell's. Malin said that one prominent neuroscientist ran an experiment with sample sizes of three. The results showed a clear difference in favor of positive transfer, but statistically significant results with such a tiny sample size are almost impossible, and the researcher concluded that because the results were not statistically significant, the reality of transfer could be dismissed.

"It was as though there was some sense of self-preservation," Malin speculated. "That he knew that he'd better *not* get a positive result."

Golub too knew of apparently positive results that were obtained in another lab, but abandoned in disbelief because "it just went against the grain of where neuroscience was at that time."

The one other lab at the University of Michigan most likely to have been interested in the transfer work was that of Bernard Agranoff. When asked years later whether he had conducted any transfer experiments, Agranoff's memory was vague. At first he said he had not, but later wrote this clarification:

> It turns out that in the very early days of our shuttle-box work, to be specific in May of 1964, we tried an experiment in which we injected RNA preps from naive and trained goldfish into naive goldfish. On the first day of testing, it looked like there might have been some increased avoidance in the fish that received the RNA from trained fish, but it was not significant. When we added additional fish, the effect went away. Later that year in August, we repeated the experiment using brain homogenates instead of RNA. It, too, was a bust.

That would have been before any attempts to transfer learning in rodents had been published, and long before the sophisticated experiments that Malin, Golub, Ungar, and a few others were carrying out in the early Seventies. Among the persevering efforts underway at that time, Ungar's was by far the most ambitious. Armed with the dark-avoidance paradigm which he felt was extremely robust and reproducible, Ungar had embarked on the attempt to chemically isolate and identify, rather than simply speculate upon, materials produced in the brain that had behavior-altering capabilities. In the heroic dimensions of his ambition and in the increasing solitude of his effort, Ungar was very much the modern day knight in search of a scientific holy grail. His admonition to Malin should more appropriately have been directed at himself.

With the material that Ungar had given him, Malin initially got positive transfer, becoming one of several labs outside of Houston to do so. With later batches of supposed scotophobin, the results were not positive, but the initial success had been sufficient to keep Malin interested. As controversy over the transfer phenomenon grew, and as the number of participants in the field diminished, it became clear that the future of the whole paradigm rested on the success of Ungar's quest. Bloodied but unbowed by his association with McConnell, Malin very much wanted to be a part of it. He was thus thrilled to be offered a postdoctoral appointment in Ungar's lab upon completion of his doctorate at Michigan.

The first few days after his arrival in Houston, Malin and Ungar had heated arguments about what Malin should work on. Malin was utterly convinced that behavioral issues—the specificity of the effect, the role of conditioned fear, and so forth—were utterly crucial to the credibility of the transfer work, and to understanding the mechanisms involved. To Malin, the psychologist, this was so obvious; but to Ungar, the years of criticism had still not impressed upon him the significance of the behavioral issues or the political impact they had had on acceptance of his research.

By persisting, Malin won a concession from Ungar to at least test for the transfer of conditioned fear hypothesis. One of the long-standing criticisms of Ungar's work was that his brain extracts simply transferred some substance produced by anxiety or fear in the donor animals, an interpretation that David De Wied gave to his own positive transfer results. Through a series of carefully controlled experiments, Malin was able to show that scotophobin *did* increase anxiety in recipients, but in an environmentally dependent way. That is, for instance, it heightened anxiety for recipients placed in a dark box, but not in a white or lighted box. He further found that several aspects of the environment influenced the transfer effect. These results were not entirely consistent with Ungar's simplistic interpretation that "fear of the dark" was being transferred; but they did support the notion that specific elements of behavioral experience had chemical consequences reflected in the population of peptides from donor brains that could influence the behavior of recipients with some degree of specificity. In a larger sense, this is all that Ungar had needed to argue, even if sometimes he seemed to go beyond that.

Ungar was pleased with Malin's work. Initially rejecting Malin's argument that behavioral questions influenced acceptance of the transfer paradigm, Ungar had given only "grudging" approval for Malin to proceed with his experiments.

"Frankly," Malin said, "I think . . . he feared that I would come up with some puzzling or negative results that would cast doubt on the whole thing. When he found that some of the results were . . . largely in agreement with his theories, then he decided this was a valuable kind of research."

But coming in the wake of the scotophobin publication debacle in *Nature*, the fine work represented in this effort went largely unappreciated, as the scientific world had concluded that the transfer work was too severely flawed to be taken seriously. Not coincidentally, most of the funding for transfer experiments seemed to dry up at this time as well. Ungar lost his long-standing NIH support. This was an irony, since the money had been provided through years of anticipation that specific molecules would be isolated and identified; then just when the goal had been achieved, financial support for the critical follow-up research was withdrawn. It was a devastating blow to Ungar, and it convinced him, in Malin's words, that "the controversy would not be settled by this generation of scientists."

Poetic Justice

It would be a stretch to suggest that the Vietnam war had a direct impact on neuroscience research. But to the extent that poetic justice is at work in the universe, the war in Vietnam was ending just as the research on endogenous opiates, which the terrible toll of drug abuse among soldiers of that war had stimulated, was reaching a climactic achievement. And this, in turn, legitimized beyond all doubt, the functional significance of peptides in the brain.

By the early Spring of 1975, labs headed by Hans Kosterlitz, Solomon Snyder, and Avram Goldstein were all closing in on identification of an endogenous opiate, which everyone agreed by then was a peptide of some sort. John Hughes, working in Scotland with Kosterlitz, had been at it the longest, but still seemed frustratingly far from identifying what was turning out to be an elusive and apparently quite small molecule. Two circumstances, however, boosted his chances of success. First, he negotiated a working agreement with Barry Morgan at the Reckitt and Colman Drug Company, whose nearby facilities could process more of the massive amounts of pig brain needed for the isolation procedures. Second, he turned to the use of high performance liquid chromatography (hplc) as an improved and much more sensitive methodology for isolating and purifying the small peptides he was looking for. This enabled him to finally start delivering samples of sufficient purity for Linda Fothergill, the biochemist who was doing amino acid analysis and sequencing for him. By the late Spring, Hughes and Fothergill knew the identity of four of the amino acids: tyrosine, phenylalanine, glycine, and methionine.

At a meeting of the International Narcotics Research Club in northern Virginia in late May, the three major labs in the quest for the chemical identity of endorphins—the generic term that would soon be adopted for all the endogenous morphine-like peptides in the brain—reported on their progress. Gavril Pasternak, working with Solomon Snyder, had confirmed the presence of one amino acid: tyrosine. Brian Cox from Avram Goldstein's lab, reported data on peptides with opiod activity from the pituitary glands which appeared to be larger and much more potent than the enkephalin of Hughes and Kosterlitz, though its chemical identity was still unknown. John Hughes appeared to be the closest to a definitive chemical characterization with his report of the four amino acids in the approximate proportion of three glycines, one phenylalanine, one methionine, and one tyrosine. The total number of amino acids, and their sequence, still eluded him, however.

It was at this point that Howard Morris, a mass spectrometry researcher at Cambridge University, entered the picture and provided a critical edge, just as Dominic Desiderio had done with Roger Guillemin. And ironically, just as Desiderio had begun his association with Ungar by attending a lecture by Ungar on scotophobin, Howard Morris first met John Hughes at a lecture by Hughes on enkephalin. This first meeting between the two at Cambridge in February of 1975 was brushed off by Hughes, who didn't have much faith in mass spectrometry at that point. But Barry Morgan was more optimistic after meeting Morris later. So with progress on the final sequence of enkephalin stymied by the summer of 1975, Hughes and Kosterlitz agreed to let Morris have a small amount of the valuable peptide for analysis by mass spectrometry.

Success was not long in coming. By mid-August, Morris had confirmed the sequence of the first four amino acids as tyrosine—glycine—glycine—phenylalanine, just as Linda Fothergill had determined. Like her, however, he had difficulty distinguishing between methionine and leucine as a fifth amino acid. Less than two weeks after he had begun, on a hunch he decided to test for the possibility that both amino acids were present, but in different shorter versions of the molecule—two variants of five amino acids each, one ending in methionine and the other in leucine—rather than a single longer peptide. With a simple chemical procedure on a new batch of enkephalin now eagerly provided by Hughes, Morris confirmed his suspicion.

A synthetic peptide with the predicted structure was prepared as quickly as possible, and by mid-October bioassay tests showed the synthetic material and the natural extract to be indistinguishable. The chemical structure of enkephalin had been discovered. With Snyder and Goldstein breathing down his neck, Kosterlitz requested priority treatment from the editor of *Nature*. The manuscript co-authored by John Hughes, Hans Kosterlitz, Linda Fothergill, Barry Morgan, Howard Morris, and Terry Smith arrived at the editorial office of *Nature* on October 28, was accepted on November 13, and published on December 18, 1975. There was no back-to-back rebuttal from a reviewer, no shadow cast over the veracity of the results, no 18-month delay. The announcement of the "Identification of two related pentapeptides from the brain with potent opiate agonist activity" was immediately embraced by the scientific world, and consumed with enthusiasm by the lay press and public, while Ungar's apparent discovery of neuroactive peptides with behavior-modifying properties receded into the recesses of an archaic footnote in the history of neuroscience. Ironically, the last three amino acids on the aminated carboxy terminal of scotophobin—glycine, glycine, tyrosine—were the first three amino acids in reverse order at the amino terminal of the enkephalins—a coincidence that went unnoticed until David Wilson pointed it out 11 years later.

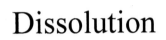

Dissolution

Rekindled Dreams and a New Strategy

The discovery of enkephalin brought Ungar's hopes briefly back to life. He, after all, had begun his whole line of research on neuroactive peptides because of the apparent discovery of an endogenous opiate antagonist in 1964 that he knew to be a peptide. Even though his years of work on the behavioral effects of peptides were being shunned, he saw in the firestorm of excitement over discovery of the endogenous opiates a vindication of his long advocacy for the importance of peptides in brain function.

David Malin likewise took heart from the new discoveries. He knew by 1974 that behavioral research on peptides, at least the work coming out of Ungar's lab, had no future, so he welcomed the redirection of Ungar's efforts toward the opiate effects of brain peptides, and joined eagerly in the venture.

About that time, Ungar visited Kosterlitz in Scotland, and returned to Houston enthused. The lab would resume its search for an endogenous opiate antagonist—a molecule manufactured by the brain whose action was mimicked by naloxone. It wasn't long before Malin and Ungar were detecting naloxone-like activity in some of their brain peptide fractions, though Malin's evidence that the peptide responsible for the effect was very small—on the order of a di- or tripeptide—was coolly received by his mentor.

"Ungar had a candidate peptide that was a little bigger at the time," according to Malin, "so he just chomped on his cigar and assumed that I had done the fractionation wrong."

"No," Malin corrected himself. "He didn't just chomp on his cigar. He lit the cigar and he blew the smoke in my face. He would do that when he disagreed with your findings."

Whatever the size of the peptide, they were soon getting clear evidence for an endogenous opiate antagonist. Malin and Ungar reasoned that the existence of such substances in the brain might account for opiate dependence—the need for ever increasing doses of exogenous opiates to produce sedation in animals chronically exposed to an opiate drug. If this were true, giving drug-dependent animals a dose of brain extract containing the opiate-antagonist peptide should precipitate withdrawal, just as naloxone would. Their initial attempts at this experiment were not successful, but then they injected the peptide material directly into the third ventricle, a region of the brain next to those known to have high concentrations of opiate receptors, and dramatic withdrawal symptoms appeared almost immediately.

Buoyed by these results, Ungar flew off to Scotland to present his work at a meeting devoted to the exciting new developments in the opiate field. But a scientific community still digesting the dramatic discovery of endogenous molecules that behave like opiate drugs was apparently not yet prepared to deal with the equally startling discovery of additional endogenous molecules that behave like *antagonists* to the opiate drugs—at least not from a laboratory already known for a previously discredited line of research on brain peptides. By all accounts, Ungar's work was received coolly at best, ridiculed at worst. He returned to Houston, according to Malin,

> . . . about as close to being crushed as somebody of his spirit—his great heart—could ever be . . . It was certainly a very sad time. Here was somebody who had worked tirelessly and with utter dedication in order to achieve certain discoveries, and it was absolutely clear that these discoveries were not going to be accepted in his lifetime.

Almost anyone else would have been overwhelmed by a sense of wasted effort, in Malin's opinion; but Ungar took the long view, feeling that someday his work would be vindicated, even if not in his lifetime.

> His ego was so strong; [because of] his own personal tendency to rely on the data, rather than to rely on opinion, I felt that he knew deep down that some of these things would be rediscovered, and would be established in one form or another, even if his name were no longer associated with them.

For Malin, the time had come to move on. His wife Jane, also a psychologist, had procured an appointment at the main campus of the University of Houston, and he was able to obtain a position at the Clear Lake campus near the Johnson Space Center south of Houston. He continued to work on the endogenous opiate antagonist, sending students to work in Ungar's lab and setting up his own lab in Clear Lake. But there ensued a period during which he could not replicate the previously robust effects that he had obtained at Baylor, and the continuing distrust of the neuroscientific community to his and Ungar's work added to his frustration.

> I don't know what the fatal attraction of these problems is for me. I think it really is much more sensible to pick tractable problems where you can make incremental advances on what has been done before. Sometimes you tilt at these windmills . . . that have defied generations of investigators, and the same thing happens to you that happened to Don Quixote—you get hoisted up on the windmill rather than solving the problem.

In a melancholy moment, Malin once told me that it had been his fortune "to work with the two all-time losers in neuroscience—James V. McConnell and Georges Ungar." Yet in the retelling of his adventures with the founder of *The Worm Runner's Digest* and the doomed discoverer of scotophobin, he betrayed no real sense of regret.

> When you've been involved in something like the memory transfer debacle, you never forget it. It leaves a mark on you. It's as though you always want to be able to come back and say, "I told you so—I was right after all." Maybe the need to be right is a strong motivation to scientists . . . I hope the need to be right doesn't become so strong that it makes us be wrong.

Having stood in his shoes, I knew exactly what he meant, and I expressed my genuine admiration for his persistence. But, he responded, "Is this a story of admirable persistence, or a story of addictive behavior?"

That, of course, is the real question. But how can we know ahead of time? Doesn't history have to decide that? If we turn out to be right, then it was obviously admirable persistence. If we turn out to be wrong, people will be free to say it was addictive behavior, and there is nothing we can do to counter it.

Are the Changes in the Numerator
or Denominator?

By August of 1973, the career of Adrian Dunn had worked its way to a rocky turning point. He had come to the United States from England to work with Edward Glassman and John Wilson at the University of North Carolina. In the early Seventies, the work of that group was generally regarded as among the best at the time on molecular mechanisms of learning. During the three years that Dunn took part in the research at Chapel Hill, much of it slowly unraveled, largely because of him and his fellow post-doctoral students, Dan and Terri Entingh. While the memory transfer work of Georges Ungar is perceived as most symptomatic of the unsuccessful efforts to make headway on learning and memory in the Sixties and Seventies, the ultimate failure of Glassman's line of research was arguably more discouraging, because for a while it had been regarded so highly.

Adrian Dunn was born in London in 1943. At an early age, he showed a ready facility for math and science—"one of those strange people . . . who worked out square roots on the backs of bus tickets at the age of eleven"—so by the time he entered Cambridge, he was well bent on a career in science, with chemistry at the center of his interest. As he progressed through his course work, he found theoretical chemistry to be increasingly esoteric and uninteresting, except for biochemistry with its applicability to biology. In 1965 he became a research (graduate) student in the lab of Asher Korner at the height of excitement over new breakthroughs in the genetic code. This was one of the few labs at the time studying protein synthesis in mammalian cells. Dunn had taken Henry McIlwain's well-known course in neurochemical techniques at the Psychiatric Institute in London, and decided to do his thesis research on the study of protein synthesis in the brain.

Restless to see the world upon completion of his doctoral work in 1968, he moved to Naples to work with Antonio Giuditta at the International Institute for Biochemistry and Biophysics. There he extended his thesis problem to the study of electroshock effects and stress on protein synthesis in the brain, and began to appreciate the more sophisticated issues regarding function of the brain that set it apart from the study of metabolism in other tissues. In late 1969, however, student and junior faculty strikes throughout Italy made scientific work nearly impossible, so he looked for another place to continue his study of functional brain metabolism. Still anxious to see more of the world, and realizing the professional value of postdoctoral experience in the United States, he elected to accept a position with John Wilson at the University of North Carolina.

As a child, John Wilson remembers seeing "gorgeous" pictures of the University of North Carolina campus at Chapel Hill brought back by his father, who was a chemist, from scientific meetings held there. In 1950, he had a chance to go to either Duke, where Philip Handler was a stern taskmaster, or to the University of North Carolina, which evoked more pleasant associations. In choosing the latter, he was one of three new faculty to join the fledgling Department of Biochemistry at the medical school that year. He had studied biochemistry at the University of Chicago prior to the onset of World War II, then pursued graduate work in organic and analytical chemistry at the University of Illinois briefly before joining Vincent Du Vigneaud, the discoverer of the structure of the posterior pituitary hormone oxytocin, among other accomplishments, at the Cornell Medical College to study the chemistry of penicillin during the war. Completing his doctorate in 1948, he was among the first to make use of radioisotopes to tag molecules so that their fate could be followed through a sequence of metabolic conversions.

During a visit to the Oak Ridge National Laboratories, where many of the early isotopes were first generated, he got the idea that if memory were so permanent, it ought to reside in some molecule that remained stable in the brain indefinitely. If that molecule were labeled with a radioactive tag, the tag should remain in the brain without being lost or converted to something else. By letting rats drink radioactive water, he reasoned that the radioactive hydrogen atoms would be interchanged among any number of metabolic intermediates until they ended up in the molecules that were repositories for memory, at which point they would cease their metabolic wandering. By identifying the molecules, the parts of the cell, and the regions of the brain where the radioactive tag ended up, he thus hoped to identify the repository of memory.

With several hundred millicuries of radioactive water from Oak Ridge, he drove back to the Chapel Hill in the summer of 1954, let his rats drink the radioactive water, then homogenized their brains in an attempt to find the ultimate destination of his radioactive tag. Crude methods for subcellular fractionation were just being developed at the time, and he interpreted his results to indicate that the label tended to end up in the cell nucleus. The one macromolecule known to be present in the nucleus and thought to be long-lived was DNA, so DNA or some related nucleic acid became his favored candidate for the role of memory molecule.

By the early Sixties, of course, others, influenced by the seminal work of Holger Hydén and generally impressed by the revolutionary discoveries on the genetic code, were thinking of nucleic acids as possible memory repositories. Among them was Edward Glassman, a New Yorker who had come to Chapel Hill and achieved considerable success as a biochemical geneticist with fruit flies. Intelligent, ambitious, and a talented facilitator, he struck Adrian Dunn as "A little bit of a megalomaniac, and a little bit of a 'good time' person. . . . He liked to have good times and be important."

Knowing of John Wilson's expertise in tracing metabolic pathways, Glassman came to Wilson and asked him how he would do an experiment to show that RNA serves as a molecular record of memory. Wilson suggested a double label experiment, whereby a precursor molecule labeled with one radioactive tag (such as carbon-14) is injected into an animal that is trained, and the same precursor labeled with a different radioactive tag (say hydrogen-3) is injected into the control animal that receives no training. The brains from the two animals are then processed together; the RNA is extracted and the ratio of carbon-14 to hydrogen-3 in the RNA fractions combined from the two animals provides a relative measure of the extent to which more RNA was synthesized in the brain of the animal receiving training than in that of the control animal.

Kurt Schlesinger, a psychologist at the University of North Carolina, designed the behavioral aspects of the experiment. He suggested a simple shock avoidance task, whereby a mouse had to jump up on a ledge in order to avoid an electric shock. The task was simple, easy to quantify, usually required only one trial, and was apparently easy to control, since the control mouse shocked in a chamber without a ledge would presumably be exposed to a similar level of stress but without the opportunity to learn the escape behavior. John Zemp, an older student who had decided to return to school for a graduate degree upon retiring from the Navy, was assigned the task of the biochemical analyses. Another student, William Boggan, carried out the behavioral aspects of the experiment. From the very beginning, all the experimental work was done solely by students—a fact that would later prove problematic when Dunn and others tried to reconstruct what had happened.

The results, published in the *Proceedings of the National Academy of Sciences* in June, 1966, were dramatic: the trained rats reportedly averaged a 64 percent increase in RNA synthesis over the untrained controls. This and follow-up publications in the ensuing months catapulted the Glassman/Wilson lab into the limelight of research on learning and memory. Notwithstanding the astounding magnitude of differences between the control and experimental animals, there were two factors that promoted the credibility of the work in the eyes of the neuroscientific community: One was the sophisticated double-label design and apparent quality of the biochemical work attributable to Wilson's reputation as a solid mainstream biochemist, and the fact that the chemical mechanisms envisioned did not violate traditional thinking about how the brain works. The other was the sophistication of the behavioral design and the intelligent controls that Kurt Schlesinger had devised, which gave the appearance that experimental mice really differed from controls only in the specificity of the training task.

Adrian Dunn, with his biochemistry background, Dan Entingh, with training in experimental psychology, and Terri Entingh, with expertise in genetics, arrived almost simultaneously at the beginning of 1970. By then, John Zemp and William Boggan were gone, as were Linda Adair and Mary Sue Coleman (later to become president of the University of Michigan), the students who had followed up the original findings.

In the beginning, the interaction between Glassman and other members of his research group—his knowledgeable and energetic postdoctoral students included—was good. To the skeptical questioning of Adrian Dunn and Dan Entingh, Glassman seemed genuinely interested in getting at the truth of the early results and why they were so difficult to replicate. In this he was hindered, though, by the fact that he had not done any of the work himself. He could not remember, for instance, on what basis data from some of the animals in the original experiment had been discarded, as a close examination of the original lab notebooks by Dan Entingh had revealed. It should be pointed out that it is neither unusual to discard data (if there is a scientific basis for doing so) nor rare for a lab director not to engage directly in the experimental work. But in this case, the two factors conspired to leave a question mark hanging over interpretation of the original findings on which Glassman's reputation in behavioral neuroscience had been built.

As designed by Wilson, the biochemical approach had involved the measurement of turnover in both RNA (the end product) and in nucleotides (the precursor molecules). Since it was the ratio of the two that was actually measured, it was possible that the turnover of either could change without the other, and subsequent experiments by the Entinghs had indicated that turnover of the nucleotides might in fact be greater than that of RNA itself. But nucleotides serve as precursors to a number of macromolecules other than RNA, so Dunn and the Entinghs began to consider the possibility that other macromolecules in the brain were being affected by behavior and siphoning nucleotides away from their destination toward RNA. Among the prime candidates for nucleotide end products other than RNA are the glycoproteins and glycolipids (since a nucleotide-carbohydrate combination molecule is used in the construction of the long carbohydrate chains in those molecules). Thus a series of experiments on the turnover of glycoproteins and glycolipids was begun as well.

Another fruitful line of research that had been initiated by John Wilson at North Carolina had to do with phosphorylation of proteins. Inspired by the pioneering work of Earl Sutherland on the role of cyclic AMP, a nucleotide involved in the addition of phosphate groups to proteins, Wilson had hypothesized that one way of altering a protein semi-permanently would be to phosphorylate it, thus conceivably turning the altered protein into something of a "memory molecule." Supposing that the relevant proteins were located in the nucleus of the nerve cell, he began to study the phosphorylation of nuclear proteins with another graduate student, Barry Machlus. Joining this effort was

Willem Gispen, a postdoc from The Netherlands, who focused on the phos-phorylation of synaptosomal proteins.

The other line of research from the North Carolina group that eventually was to have lasting significance was a continuation of work begun by Dunn in England, and independently by Gispen in the Netherlands. Each had shown that ACTH, the pituitary hormone that stimulates the adrenal gland's response to stress, could promote protein synthesis. In North Carolina they learned that Lynda Uphouse, one of Kurt Schlesinger's graduate students in psychology, had found that animals undergoing behavioral training showed an increase in protein synthesis, though it occurred to some degree in the liver as well as the brain, and in the yoked (stimulated but untrained) as well as in the trained animals. For these reasons, Glassman had shown little interest in the findings. But to Dunn and Gispen, it immediately suggested a non-specific stress response. And in view of the findings of David De Wied in the Netherlands that ACTH improves the ability of animals to learn, a logical connection between stress, protein syn-thesis, and learning was immediately drawn:

1. Training enhances protein synthesis.
2. ACTH promotes protein synthesis.
3. ACTH facilitates learning.
4. The stress of training promotes the secretion of ACTH.
5. Therefore, stress-induced secretion of ACTH promotes learning by stim-ulating protein synthesis.

"These kinds of things occur to you in about half a second—the whole pattern," Dunn recounted. "So that is how I started studying ACTH and protein synthesis."

In the summer of 1972, the Neurosciences Research Program held its third Intensive Study Program in Boulder, Colorado. As before, the brightest and best of the world's young, emerging neuroscientists (which necessarily meant those associated with the best known senior neuroscientists) were invited to attend. Glassman was allotted the privilege of bringing one postdoc from his group, and Adrian Dunn was selected for the honor by a flip of the coin (or some other form of chance selection that Dunn does not recall). A new innovation this time was the inclusion of brief chapters authored by the junior scientists themselves in the large tome that emerged from the meeting. Thus, the volume from the Third Study Program bears not only a chapter by Glassman, but another with A. Dunn, D. Entingh, T. Entingh, W. Gispen, B. Machlus, R. Perumal, H. Rees, and L. Brogan as co-authors.

The chapter by the postdocs and graduate students of Glassman and Wilson painted a skeletal picture of the various chemical directions into which their research had diverged. Included were data on phosphorylation of nuclear pro-teins, phosphorylation of synaptic proteins, turnover of glycoproteins and gang-liosides, and amino acid incorporation into proteins. Though constituting a frustrating litany of approaches to those who sought a more straightforward and simplistic chemical correlate of memory storage, the multifocal strategy made a lot of sense, because, in the words of the authors,

> . . . learning is a complex process, and it is thought likely that particular components of the training situation, such as stress, arousal, attention, motivation . . . may in fact be necessary aspects of the learning process.

In 1969, Glassman had published a review of protein and nucleic acid changes in learning, which though widely heralded at the time, was characterized by the theoretical simplicity that typified nearly all new entrants into the field whose experience was long on biochemistry but short on brain function. By contrast, Glassman's chapter in the Study Program volume in 1972 was much more insightful and theoretically sophisticated. For instance, by now he had come to an understanding that behavioral controls for the specificity of the information acquired in a learning task are virtually unattainable. He pointed to "exaggerated concerns" about whether biochemical changes in a learning experiment are specific or non-specific to the information learned, noting

> The difficulty, mostly semantic, is that nonspecific responses do not exist; they are merely responses whose cause is unknown. It would probably be more productive to define such responses in definite terms . . . stress-specific, attention-specific, performance-specific, etc.

Much of the chapter was devoted to an elaborate hypothetical scheme of how a large number of biochemical reactions might participate in an interrelated matrix of events leading to either short-term or long-term changes in synaptic efficacy. In this respect, Glassman was clearly reflecting the influence of John Wilson and his younger colleagues, which he acknowledged in singling out Dan Entingh, Adrian Dunn, and John Wilson for special mention.

Those two chapters in 1972 probably reflected the high-water mark for the group at North Carolina that Glassman, Wilson, and Schlesinger had put together. Some of the individual leads would be followed up and published in detail, but the picture became ever more complex, plagued by the variability that seemed to afflict all experiments of that type at the time. Internal second-guessing of the data on which the original claims for changes in RNA associated with learning had been made, detracted from the luster of those findings, and painted Glassman into an increasingly defensive posture. "The more that Glassman came under this onslaught," Dunn observed, "he simply didn't want to face it. He just sort of tried to stay above it, and ignore it, and that became frustrating . . . he became more and more reclusive . . . and never recovered from it."

Glassman had been the glue that held the group together and the catalyst that energized it. Therefore, his retrenchment made dissolution of the group inevitable. In time, his relationship with Dunn turned hostile. Realizing that there was no future for him at North Carolina, Dunn accepted a faculty position at the University of Florida, where he continued to pursue his highly fruitful line of research on stress-induced behavior mediated by various endocrine events, helping to clarify some important issues in behavioral biochemistry. Willem Gispen returned to Europe, married David De Wied's daughter, and continued to make seminal contributions in the field of protein phosphorylation. John Wilson maintained the basic neurochemical work on the phosphorylation phenomenon in the same lab that had known more exciting times. But the early optimism of the original work that had seemed to point to RNA as a molecular repository of memory faded into the recesses of unproven speculation.

Downward Spiral

As the story of scotophobin played out and the limelight shifted finally and totally away from RNA as a memory molecule, I was preoccupied with professional survival. The decision to go to New York had been personally satisfying, and it gave Carol and me an experience that we would never regret. But it was not a wise move for my career.

The College of Pharmaceutical Sciences—Columbia University, as it was officially named, was associated with Columbia in name only. Financially independent and in competition with the recently established state university system of New York, it had begun to lose enrollment and spiral into an irrecoverable deficit. It was not a state-of-the-art facility to begin with, and its financial troubles quickly made support for research totally dissipate. I managed to obtain a small grant that generated research for a couple of brief reports on sialic acid metabolism in brain tissue, but heavy teaching and institutional turmoil took their toll on my productivity. Thus, when it became obvious that, as the last faculty member hired I would be the first casualty of the institution's financial crisis, it was probably a blessing for me to seek a position elsewhere.

Through Jerry Mitchell, a close friend from graduate school who had become an Assistant Professor of Anatomy at the Wayne State University School of Medicine in Detroit, I procured an interview in the Department of Physiology. That Department was world-famous for its research on blood coagulation—a subject I knew almost nothing about—but it was looking for a neurochemist to bolster its meager program in neuroscience, so my timing was good. I was offered the position and moved thankfully to Detroit in the summer of 1973.

There I met Robin Barraco, one of the most dynamic and stimulating colleagues I would ever know. An Assistant Professor of Physiology hired only a year or two ahead of me, he and I became fast friends and close collaborators, studying a whole host of biochemical changes in the brains of operantly conditioned pigeons. Except for our first paper together, in which we demonstrated the presence of free peptides in the pigeon brain, he and I both stayed away from neuroactive peptides, content to watch the drama unfold in that field from the sidelines.

My time in Detroit didn't last, however. Robin and I worked hard and published several papers during my three years at Wayne State, but we had not managed to get a grant by the time I had to be considered for tenure, and that insufficiency left me looking for a new position again. Having spent a previous summer at the Neurosciences Research Program (NRP) in Boston, my inquiry was welcomed and I flew to Boston for an interview in April, 1976. Somewhat to my surprise, I was welcomed by Frank Schmitt like a member of the family, and offered the position of Staff Scientist on the spot. I remained at Wayne State for a final semester in the fall. Ironically, Robin and I then got our first grant— an NSF award to study cell-surface glycoproteins and glycolipids during neural development. When Wayne State declined my invitation to reconsider its rash decision to deny me tenure, Carol, our son Anthony, and I moved to Boston in late December, where I assumed a position with responsibilities that were vague at best.

The job description included support for NRP's ongoing program of work sessions on current hot topics in neuroscience. What exactly the Staff Scientist at NRP was supposed to do was never explained with clarity. There were vague references to something called "conceptual research," but what that entailed was totally unclear. Adding to my personal frustration was one of the coldest winters in my experience. With work responsibilities sufficiently undefined, and surviving the daily commute through the twisted and snow-choked streets of suburban Boston as my primary daily challenge, about all I could concentrate on during the first year of my return to the east coast was writing up left-over research from Detroit and resuming some focus on what had happened to scotophobin.

During this period of struggle for me with the harsh weather of New England and the puzzling demands of conceptual research at NRP, Georges Ungar was struggling with a more serious crisis. He had contracted cancer and was gravely ill.

The previous summer, his funding had finally been exhausted and his lab space at the Baylor College of Medicine was lost when he chose the wrong side of the power struggle between heart surgeons Michael De Bakey and Denton Cooley. So he and Alberte had moved to Memphis, where Bill Byrne was able to provide Ungar with a title and a little space but no salary. Alberte was also given a technical position with a little bit of salary. Ungar knew that the end of his career was looming, but he had no intention of fading away quietly. Indeed, he looked forward to doing the more philosophical and speculative writing that had always appealed to him, but for which he had seldom had time in the past.

Still honored in Europe to a greater degree than in the United States, he had been invited to participate in a meeting in Leningrad. He and Alberte had flown there in early November, and, after the meeting, had taken a side trip to Paris to enjoy the city of their youth. It was there that the symptoms of the cancer first manifested themselves in force, though in a misleading way. Gastrointestinal upset was the main malady; and by the time they got back to Memphis, Ungar knew he had to see a doctor immediately.

The comprehensive work-up included a chest x-ray, which revealed a large tumor at a branch point in one of the airways of the lung. Probably it had been secreting gastrointestinal hormones, which caused the stomach upsets, as these types of tumors sometimes do. Analysis showed it to be a spiculated cell carcinoma, fast evolving. This type of tumor can spread quickly to other organs, especially to the brain from the lung. Ungar knew this, and knew the prognosis was grave. Lung carcinomas are sensitive to radiation, though, and at Bill Byrne's urging, the Ungars decided to seek treatment at the M. D. Anderson Hospital in Houston, the city where they had been most content, and where they still had supportive friends. In mid-December of 1976, an aggressive regimen of chemotherapy and radiation was begun, which initially dissolved the main tumor. This treatment continued on a monthly basis through the spring of 1977.

In retrospect, Alberte thought that seeking treatment at M. D. Anderson had probably been a mistake. The waiting, often for hours, seemed interminable to Ungar, and his former position as a distinguished faculty member at the Baylor College of Medicine earned him little extra sympathy. Even though acquaintances of his on the medical staff at M. D. Anderson *did* extend him as much special attention as they could, Alberte recalled that

> Georges had never been really ill in his life, and he took very badly to being uncomfortable. He was a very poor patient, Georges. He was not trying to get over it. He thought he was finished, and that was that

Nonetheless, the success of the early treatments invigorated him to a degree, so he made plans to go to the American Society for Neurochemistry meeting in Denver in March, 1977. His daughter Catherine would join her parents there, and they would make a little mini-vacation of it. I too had long planned to attend this meeting—especially because it was in Denver, but also because two of the students from our lab at Wayne State, Dave Terrian and Marc Abel, would be giving their first papers at a national meeting there.

On Wednesday evening, I met with Georges, Alberte, and Catherine for cocktails and a nice visit. The moment I saw him, I could tell he was very ill— thin and gaunt with all remnants of hair gone. But he was in good spirits and appeared genuinely happy to see me. When he heard that I had been awarded an NSF grant and would be restarting my research at the Eunice Kennedy Shriver Center that summer, he was particularly pleased. Later, after he had excused himself in order to rest before a panel discussion in which he was later scheduled to take part, Alberte filled me in on details of his condition, adding, "As you know, he is very fond of you."

The panel discussion provided a sad counterpoint to the happier social hour earlier. Billed as a panel discussion on the biochemistry of memory transfer, the program consisted of a collection of spokesmen for novel (some would say fringe) approaches to the study of biochemical correlates of learning. Samuel Bogoch talked about his controversial brain glycoprotein studies, Ungar defended the behavioral transfer experiments as a form of bioassay, Barry Kaplan talked about changes in nucleic acids associated with experience, and J. L. Sirlin spoke on changes in transfer RNA with training in goldfish. Detractors in the audience were not in a compromising mood. It was clear from the comments of several younger, aggressive attendees that they had no understanding of the history of the field, though they hardly let that inhibit the expression of their strongly held, uninformed opinions. As usual, Bogoch was a lightning rod, drawing the ire of Richard Margolis, whose disagreements with him over technical issues degenerated into a shouting match. Carol had worked for Margolis when we lived in New York, so he and I knew one another personally as well as professionally. When Bogoch invoked the name of Barraco and Irwin a couple of times in support of his own work I cringed a little, though more in embarrassment for Carol, as I could imagine Margolis wondering how his respected postdoctoral student could have married a person prone to get mixed up with characters like Bogoch. The fact of the matter was, however, that neither Margolis nor anyone other than Robin and I had attempted to replicate Bogoch's provocative results. It saddened but no longer surprised me to see how emotional and impassioned attitudes about this type of research had become.

Two evenings earlier, Adrian Dunn had chaired a round-table discussion that had barely fared better. In his continuing effort to decipher the role that peptide hormones play in modulating memory storage and retrieval processes, he had attempted to lead a discussion that looked at these compounds in that new light. But the relationship of peptides in any form to behavior in any way had become so tainted that the audience as a whole was clearly unreceptive. This struck me as all the more strange in the light of the opiate peptide mania that was sweeping the world of neuroscience at the time. But in fact, the opiate craze had brought a lot of new, young people into neuroscience who had little inkling of what had gone before; and those who *had* bothered to read the literature could not have failed to be impressed with the apparent disarray into which the field of peptide research had fallen.

I returned to Boston, to the privilege and frustration of trying to define and accomplish something meaningful at NRP. But much on my mind was the flurry of work and excitement on neuroactive peptides that discovery of the opiate receptors had generated, and the melancholy realization that for Georges Ungar, the vindication was too little, too late, and not quite close enough to the thrust of his own work to bring him the credit that he surely longed for.

One day in early Spring, Alberte returned from a meeting to find Georges sitting at his desk, staring blankly, and mumbling incoherently about a drink he had spilled on himself. Fearing a stroke, Alberte rushed him to the emergency room, then watched him intently over the next 24 hours, as he seemed to pull out of whatever had afflicted him. Over the ensuing days, he regained his faculties;

but the incident was clear evidence that neurological symptoms from the metastasized tumor had set in.

Later in the spring, a visit from Jay Nutt, one of the students who had worked with him during the heady days of early success at Baylor, cheered him. But the cancer had spread too far too fast, and even aggressive treatments were no longer sufficient to keep it in check. Because food was so unpalatable, Ungar ate almost nothing, and weakened quickly.

Catherine had joined her parents in Memphis, to help her mother take care of him. As he weakened, the neurological symptoms increased. It was the latter that precipitated one of the final fights of his life. Catherine and her mother were trying to change the sheets on Georges' bed with him in it one day, when he apparently felt himself under attack. Mounting a ferocious defense, he fended off his assailants with all the meager strength left in his dissipated body. Taken by surprise, the two women found themselves in an intense three-way wrestling match that led to their collapse in disarray, then in long pent-up and bitter-sweet laughter over the implausible scene, as Ungar stared at them with fire in his uncomprehending eyes. It would be the last spark he would ever show.

Ungar and Samson

In the twilight of Georges Ungar's life, when hundreds of young scientists who never heard of him and scores of older scientists who knew his work but seldom cited him were working on endogenous peptides with opiate activity, Ungar returned to the finding he had made a decade before endogenous opiates were discovered—namely, the apparent presence in brain of naturally occurring peptides with *anti*-opiate activity.

In publications at the end of 1976 and the beginning of 1977, Malin and the Ungars published evidence that chronic injections of morphine induced the appearance of a host of peptides in the brains of rats, some of which acted as opiates, and others of which appeared to be opiate antagonists. One such peptide, which had the structure of met-enkephalin with arginine added at the end opposite to methionine, mimicked the effects of met-enkephalin in the vas deferens bioassay, but acted like an opiate antagonist when injected into an intact animal. Their experiments suggested that exposure to opiate drugs induces the production of both agonists and antagonists, both of which fall off in time (though the agonist declines faster than the antagonist, suggesting a mechanism for the development of drug tolerance).

Even these exciting findings, though now well within the theoretical framework embraced by most neuroscientists in the wake of the discovery of the endogenous opiates, was met by what Malin described as "a world-wide outbreak of indifference." The neuroscientific community, as it plunged frenetically forward in quest of ever more intricate details about opiate receptor kinetics and distribution, was impatient with attempts to look at all the neuroactive peptides as a family of complexly related substances. That time would come, but in early 1977, it had not yet arrived.

For Georges Ungar, it mattered little as spring turned to summer. Unable to eat and consumed by cancer, he deteriorated rapidly. Two rounds of chemotherapy had held the tumors in abeyance, but the third round had failed to do so.

When it became apparent that the end was near, Alberte had him moved home from the hospital, and called Catherine to his bedside. Through the night of July 25 they kept a vigil. He slept peacefully and without apparent pain. About 7:00 in the morning of the 26th, he took a few deep breaths—"the way animals die," Alberte observed—and stopped breathing. At the age of 71, his life had come to an end.

Bill Byrne called me from Memphis that afternoon with the news. I had known that his disease was terminal, and had not really expected to see him again after the meeting in Denver the previous March, but news of his death still depressed me deeply. Whatever his failings, to me he had been a friend from the beginning. His faith in my promise from that first day we met in December of 1965 had touched me at a fragile time in my life, and my work with him— whatever its ultimate worth—had provided me with the profoundest experience of my scientific life. From a purely egocentric perspective, I felt that I had lost one of my strongest champions.

David Malin delivered the eulogy at his memorial service, expressing eloquently many of the feelings I shared:

> On this sad occasion, I would like to do my best to strike a happy chord. For today we celebrate the life of a happy man. He was happy to spend most of his life doing exactly what he enjoyed most— patiently asking questions of nature in his laboratory and occasionally receiving her answers. He was happy to share his deep interests with other scientists who were not just colleagues, but friends and comrades. He was uniquely happy to share these interests with the person closest to him, his wife Alberte, as they worked side by side. He was happy to see his daughter, Catherine, grow up absorbing and reflecting his love of scholarship. . .
>
> It was characteristic of Georges that he always pursued lines of research that were both original and important. He never simply hopped on the bandwagon and did the kind of experiments that were popular or easy at any particular time . . .
>
> We all know that Georges' striking later discoveries were controversial and not accepted at all in some quarters. This didn't crush him as it would some others, because he was interested in data, not popularity; in logic, not scientific politics. He was confident that the self-correcting nature of the scientific method would result eventually in a fair assessment of his theories and, more importantly, in a correct understanding of the physical basis of mind.
>
> Finally, Georges .   . demonstrated courage by proceeding calmly with his work and his ideas despite harsh criticism and misunderstanding. . . . His final gift to us is a lesson in how to face death with courage. Yet, it was really a lesson in how to live – how to live a life so full that, when the time comes, one will be able to face its conclusion with no regrets.

In the years that followed the death of Georges Ungar, some remarkable advances would be made in understanding the cellular basis of neural plasticity.

And neuroscience would continue to expand in scope and break into ever more specific subdisciplines. It would become institutionalized in academic departments, in places like the University of California at Irvine and the University of Texas Medical School in Houston. As in the maturation of all fields, however, the growth and institutionalization brought fractionation, reductionism, and a consolidation of prevailing views. With the death of Georges Ungar, one of the last of the classic scientists to pose an overarching link from biochemistry to behavior was gone. The scientific triumphs of neuroscience would march forward, but the early age of neuroscience, when it was possible to propose a unified if inadequate explanation for the molecular basis of memory, died with the last of the scientists who had tried to do so.

If Georges Ungar had been my hero, Fred Samson had been no less so. He would live on for more than two decades. In 1972, he had moved from the campus in Lawrence to the Ralph Smith Center for Mental Retardation at the medical campus of the University in Kansas City. In 1990, he stepped down from that position, demonstrating at his retirement dinner the one-armed hand stand he had perfected as an acrobat in his youth. He retained his office and continued to work vigorously every day, in concert with Frank Schmitt until Schmitt's death in 1995, then on his own, pursuing especially questions related to the role of free radicals in biological systems. He never got over his anger at Ungar, and never believed that Ungar had been less than a fraud. After a prolonged illness that left him incapacitated for the first time in his life, Fred Samson died on April 15, 2004.

The End

Addendum

Historical Context of Research Attempts to Transfer Learning Between Animals by Biochemical Means

In the previous pages, I sought to relate in narrative form the background and circumstances that led up to and followed the characterization of scotophobin, a neuropeptide reputed to encode dark avoidance behavior in rodents. It was intended as a personal account of my perceptions, subject to what I knew at the time and have since learned. To help the story flow smoothly for the general reader, I avoided inserting citations to the published record of the scientists whose work I described. In the hope of giving this treatise more historical value, however, I will now provide that documentation in the historical essay that follows. Research on learning and memory has been, of course, much broader than the approach reviewed here. The following account, therefore, is merely intended to document the endeavors referred to in the earlier narrative portion of the book, rather than provide a comprehensive overview of research on the molecular basis of memory.

By the start of the 20th Century, the conspicuous success of theoretical physics had confirmed the utility of a mechanistic view of the natural world. This provided an acceptable framework for students of the brain, impressed by the beautiful structural details revealed by Ramón y Cajal (1899), the mechanical nature of reflexes (Sherrington, 1906), and the determinism inherent in Pavlovian conditioning (Pavlov, 1927), to think of brain function in structural, mechanistic terms. Even the amorphous psychoanalytical insights of Sigmund Freud found expression in structural schemata (Freud, 1900). Increasingly, the brain became envisioned as a network of finite albeit incredibly complex pathways (DeCamp, 1915; Hering, 1895; Konorski, 1950; Sherrington, 1906). The development of telephone switchboards, and later computers, encouraged this view (Adey, 1969; Rose, 1998).

In psychology, the stimulus-response-reward theories of Thorndike, Hull, Guthrie, and Skinner promoted a robust connectionism that influenced generations of brain theorists (Hilgard, 1956). The notion of the brain as a hard-wired network became fixed, though it was appreciated that the modifiability of behavior implied the necessity for a mechanism of plasticity within the network. Ideas of how this might occur go back to the turn of the century (DeCamp, 1915; Freud, 1900; Thorndike, 1911), but reached their most compelling and influential form in 1949 with the publication of Hebb's *The Organization of Behavior*. Hebb's "neuropsychological theory" was mostly a sophisticated attempt to explain cognitive science in mechanistic terms, but his enduring influence on thoughts about brain plasticity derived from his clear and succinct proposition that "When an axon of cell A is near enough to excite a cell B and repeatedly or persistently takes part in firing it, some growth process or metabolic change takes place in one or both cells such that A's efficiency, as one of the cells firing B, is increased." (p. 62)

At mid-century, connectionist views of brain function were augmented by the discovery of the bioelectrical basis of the action potential (Hodgkin and Huxley, 1952), and the nature of chemical transmission (Eccles, 1961; Eccles, Eccles & Fatt, 1956), again pointing to information transaction through fixed pathways, possibly subject to fairly simple chemical modulation at the nerve ending (Eccles, 1977; Eccles & McIntyre, 1953; Nelson, 1967). Learning theorists were inhibited from adopting an unabashedly hard-wired model of neural networks by the difficulty of pinpointing the location of specific memories. While the loss of certain specialized motor and sensory functions could be attributed to specific regions of the cortex, and while direct stimulation of the brain could unleash the recall of specific trains of memory (Penfield, 1955; Perot & Penfield, 1960), the distributed properties of cortical activity, and the propensity for "equipotentiality" in cortical function were compelling (John, 1961; John, 1972). The fact that information degradation following extirpation of cortical tissue was tied more closely to the amount of tissue removed, than to the precise locus of deletion, was most elegantly demonstrated in a series of studies by Karl Lashley (1929), culminating in his frustrated rhetorical suggestion that memory might be an impossibility (Lashley, 1950). The inability to localize specific memories (engrams) left the door open for those in search of an alternative to the growing sterility of the connectionist model of information in the brain (Hechter, 1966; John, 1961).

That alternative was soon to be sought by analogy with the seemingly inscrutable mystery of hereditary coding of information, the ultimate problem in bioinformatics. By 1950 it was accepted that DNA was the genetic material, though how a molecule of such apparent simplicity could code information of such complexity was a severe problem (Judson, 1979). With the storage of hereditary information being the ultimate mystery, it was inevitable that its solution would have repercussion on ideas about the storage of behavioral information.

Notwithstanding the prevalence of connectionist notions about neural networks, some biochemically-oriented scientists had started to look for signs of chemical plasticity by the 1940s (Halstead, 1951; Monné, 1948). Holger Hydén was notable among them, providing evidence as early as 1943 that sustained stimulation of nerve cells increased their staining properties, suggesting the elaboration of macromolecules, including nucleic acids, consequent to increased functional activity (Hydén, 1943). This work persisted into the early 1960s, with the added aura provided by his elegant microchemical analytical techniques, in which he analyzed the nucleotide base ratios in the nucleic acids from individual cells (Hydén, 1959; Hydén, 1960; Hydén, 1966).

In 1953, the structure of DNA was solved, suggesting immediately a replicative process and disclosing the fact that information must be stored in the sequence of nucleotide bases in the DNA strand (Watson & Crick, 1953; Wilkins, Stokes & Wilson, 1953; Franklin & Gosling, 1953). This fact was taken up almost immediately as a paradigm for the storage of acquired information in the brain. If information resides in the sequence of nucleotide bases, a change in the relative composition of different bases would automatically imply an altered sequence, and altered information content. This was the theoretical rationale for Hydén's (1959) search for base changes in nucleic acids associated with learning, and as offered in evidence by the experiment in which RNA from the brainstem of rats was shown to change as they learned a difficult balancing skill (Hydén & Egyházi, 1962; Hydén & Egyházi, 1963). A similarly elegant experiment appeared to produce similar results when rats were trained to reach for food with one paw but not the other (Hydén & Egyházi, 1964).

By this time, the search for chemical changes was focusing on RNA, since it was conceded that DNA must be the unalterable repository of genetic information. RNA on the other hand, was produced anew with each cell cycle and was presumably capable of being modified, on the one hand, and perpetuating the modification on the other. How either was accomplished was the subject of pure speculation, since the precise role of RNA remained obscure through the re-mainder of the 1950s, and some skeptics warned about an overly facile com-parison between genetic and experiential information storage in nucleic acids (Briggs, 1962; Gaito, 1963; Goldberg, 1964). It was not until the discovery of messenger RNA (Brenner, Jacob & Meselson, 1961; Gros et al., 1961) and the basic mechanisms of gene regulation (Jacob & Monod, 1961) that a clear understanding of the flow of information from the genome (DNA) to expression of the gene product (protein) was clarified. Then, not until the advent of cell-free translational systems made elucidation of the genetic code feasible (Nirenberg et al., 1963; Nirenberg & Matthaei, 1961), was it possible to reconstruct the sem-antic structure of genetic information storage at the molecular level.

Thompson and McConnell (1955) began their experiments with flatworms at the height of excitement over discovery of the structure of DNA and at a time of growing appreciation for the informational role of RNA, but prior to a clear understanding of how, precisely, nucleic acids decode the information contained in their base sequences. Aware as he was of Hydén's work, it was natural for McConnell to look to RNA as a repository of behaviorally acquired information. McConnell was by no means the only one to do so. The literature of the period is replete with a lot of speculation and some experimentation on the role of RNA as repositories of the engram (Corning & John, 1961; Gaito, 1963; Hydén, 1959; Mihailovic et al., 1958). Thus, the early emphasis on RNA as potential memory molecules makes sense, in light of the level of uncertainty at the time over information storage and readout at the molecular level.

Holger Hydén's work continued to be influential, as he kept up a line of research on changes in nucleic acids associated with learning, and adjusted his theoretical explanations to be in tune with the progress made toward a clearer understanding of information transaction at the molecular level (Hydén, 1959; Hydén, 1967; Hydén, 1970; Hydén, 1976; Hydén & Lange, 1965; Hydén, Lange & Perrin, 1977). By contrast, McConnell's career took on a show-business character that he did not discourage. By nature, it was impossible for him to avoid the inherent humor engendered by the cannibalism experiments with flatworms, in which he claimed to demonstrate the transfer of information to "recipients" from ingested "donors" (McConnell, 1962). The carnival-like aspects of the experiments, and McConnell's own tendency to see and even highlight the implicit humor in them, obscured what in a sense was a boldly creative experiment carried out and published as a serious scientific endeavor by numerous investigators (Corning, 1964; Corning, 1966; Corning & John, 1961; Jacobson, Fried & Horowitz, 1966c; McConnell, 1962; McConnell, Jacobson & Kimble., 1959; Thompson & McConnell, 1955; Westerman, 1963), including Hydén himself (Hydén, Egyházi & John, 1969). To be sure, there was controversy from the beginning about the reliability of planarian conditioning (Bennett & Calvin, 1964). Also, because some of the most startling claims of transfer in rodents would come later from McConnell's students and associates (Babich, Jacobson & Bubash, 1965a; Babich et al., 1965b; Jacobson et al., 1966a; Jacobson et al., 1966b), and because many of those later investigating the transfer phenomenon drew their inspiration from the cannibalism experiments with planaria, they and the transfer experiments in vertebrates became attached in the minds of the public and most scientists as a conceptual continuum, which did not help their overall credibility.

Before the work on planaria had begun, Mark Rosenzweig and David Krech, two psychologists at the University of California at Berkeley, had teamed up to test the hypothesis that differences in synaptic neurochemistry could explain the tendency for rats to use different strategies in learning a maze. They were joined in this effort in 1953 by Edward Bennett, a biochemist from the lab of Melvin Calvin. Their early results appeared to confirm that differences in learning ability could be correlated with differences in the levels of cholinesterase activity in selected regions of the cortex (Krech et al., 1954; Rosenzweig, Krech & Bennett, 1958). They decided to try a relatively mild form of intervention, by exposing some rats to a more socially and psychologically enriching environment, and, in response to an increased understanding of synaptic chemistry, to distinguish between the acetylcholinesterase restricted to neural tissue from the more generalized activity of cholinesterase (Krech, Rosenzweig & Bennett, 1960; Rosenzweig, Krech & Bennett, 1960). Their need to divide by tissue weight in order to calculate specific activities soon led to the discovery that differences in cortical thickness were present as well (Rosenzweig et al., 1962). A neuroanatomist, Marion C. Diamond, joined the team, adding to their ability to make precise, regionally-specific quantitative measures of anatomical changes (Bennett et al., 1964; Diamond, Krech & Rosenzweig, 1964; Rosenzweig, Bennett & Krech, 1964). The brain was seen to be plastic after all, as everyone had known it must be at some level. That it was plastic at such a relatively gross level was somewhat surprising, but not to an alarming degree, since an integration of Hebbian synaptic modifications, either structurally or neurochemically or both, over large areas of the brain could well lead to such changes.

By the 1960s, with the glamour of biochemistry well established and neurochemistry gaining a scientific foothold, antimetabolic drugs had replaced surgical extirpation as the method of choice in searching for the substrates of memory. If learning required the elaboration of some material that altered the behavior of nerve cells, interference with the brain's ability to synthesize proteins should inhibit the animal's ability to learn. Josefa and Louis Flexner at the University of Pennsylvania initiated the use of puromycin in such an attempt (Flexner et al., 1962). Their experiments disclosed a regional specificity to the deposition of memory that appeared to broaden with time (Flexner, Flexner & Stellar, 1963), the first strong support for the notion—long held by experimental psychologists—that memory goes through different stages of encoding before achieving its semi-permanent storage form. Protein synthesis inhibition had to be profound to inhibit memory, however, and not all protein synthesis inhibitors were effective (Flexner et al., 1964). Furthermore, puromycin was shown to be a potent inhibitor of cyclic-AMP phosphodiesterase (Appleman & Kemp, 1966) as well as a protein synthesis blocker, and to cause occult seizures (Cohen & Barondes, 1967). Thus, the mode of memory inhibition by puromycin remained problematic, but the work of the Flexners accelerated momentum toward the search for chemical as well as structural substrates of memory.

Bernard Agranoff added an important cross-species perspective by inducing amnesia in goldfish. He demonstrated a gradient in retention proportional to the time following training in which either electrocortical shock (ECS) or puromycin was administered (Agranoff, 1965; Agranoff, Davis & Brink, 1965; Agranoff & Klinger, 1964). This provided a direct connection between bioelectrical and metabolic activity in the brain, and solidified the conviction that protein synthesis is necessary for memory storage. It turned out that peptidyl-protein fragments generated by treatment with puromycin were long-lasting in the brain (Flexner & Flexner, 1968). This and the other problems with puromycin noted above led Agranoff to turn to acetoxycycloheximide, a protein synthesis inhibitor with fewer side effects than puromycin. He was then able to show with greater confidence that amnesia in goldfish indeed appeared to have an absolute requirement for the elaboration of new protein (Agranoff, Davis & Brink, 1966).

Given the focus on nucleic acids as informational reservoirs in the early 1960s, and the impetus of Hydén's experiments at the time, it was natural that some of the earliest experiments on metabolic disruption would target the nucleic acids rather than protein. Samuel Barondes first took this approach, leveraging his fortuitous introduction to molecular biology (Nirenberg et al., 1963) into a direct application to brain function. Inspired by the work of Dingman & Sporn (1961), who had shown that 8-azaguanine, an abnormal RNA precursor, suppressed the ability of rats to learn a water maze, Barondes and Jarvik conducted similar experiments with actinomycin D, which blocks transcription of RNA (Barondes & Jarvik, 1964). Though their results were equivocal, they provided Barondes with a platform for thinking and speaking about the line of research attempting to link molecular mechanisms and memory. One such talk, at the 1964 meeting of the American Psychological Association, provided the basis for one of the most comprehensive and insightful reviews of the field up to that time (Barondes, 1965).

Agranoff also tried with somewhat greater success to use actinomycin D in goldfish (Agranoff et al., 1967), but the toxicity of the drug was obviously a severe limitation. So Barondes turned to the use of puromycin, as the Flexners and Agranoff had already done, but with a difference—he injected the puromycin prior to training so that protein synthesis would be inhibited *during* the learning experience. He found that rats could learn a Y-maze with protein synthesis severely inhibited, but could not recall the correct response hours later (Barondes & Cohen, 1966). This and a number of appropriate control experiments established that initial acquisition and even early retrieval (short-term memory, STM) is not dependent on protein synthesis, whereas later retrieval (long-term memory, LTM) clearly is. He and Cohen further showed that arousal contributes to the conversion of STM into LTM (Barondes & Cohen, 1968b), consistent with the growing evidence that adrenocortical stimulation facilitates learning (Bohus, Nyakas & Endröczi, 1968; De Wied, 1965).

Another of Mark Rosenzweig's students, James McGaugh, pursued the metabolic study of memory from the other direction—using drugs to enhance performance of learned behavior. In a long series of experiments, he showed that strychnine and other central nervous system stimulants led to faster acquisition and longer retention (Breen & McGaugh, 1961; McGaugh, 1961; McGaugh & Petrinovich, 1959; McGaugh & Thompson, 1962; McGaugh, Westbrook & Thompson, 1962). These experiments pointed to the dynamic nature of the consolidation process, occurring early and quite possibly independent of macromolecular changes.

The pharmacological approach to the study of learning and memory filled in vital details about the time course of consolidation and the intricacies of retrieval. While it remains difficult to this day to distinguish whether disruption of memory is due to a failure of consolidation or retrieval, these experiments, at least, helped crystallize the questions and added a lot of phenomenological detail. Because they fell well within the framework of canonical assumptions about how the brain works—as a hard-wired system of connections modifiable through chemical and possibly microanatomical changes at Hebbian synapses—they raised no fundamental conceptual alarms. The fate of the chemical transfer experiments would be otherwise.

Between April and August of 1965, four labs published claims of inter-animal transfer of behavioral tendencies in rodents by injection of brain extracts. In April, Fjerdingstad and his colleagues in Copenhagen reported that RNA extracted from brains of rats trained to approach the lighted side of a partition for a water reward, promoted the same tendency in recipients (Fjerdingstad, Nissen & Roigaard-Petersen, 1965; Nissen, Roigaard-Petersen & Fjerdingstad, 1965). In June, Reinis (1965) claimed to have increased the rate of alimentary conditioning in rats by injecting brain homogenates from rats previously conditioned in the same way. In July, Ungar (1965) reported that intraperitoneal injection of a chymotrypsin-labile but not ribonuclease-labile extract from brains of rats subjected to extinction of their startle reflex to a loud sound would accelerate the rate of extinction to the same stimulus in recipients. In August, Allan Jacobson and his colleagues (Babich et al., 1965a) claimed that RNA extracts from brains of rats trained to approach a food cup would increase the same tendency in recipients. The same group also reported cross-species transfer of learning from hamsters to rats (Babich et al., 1965b).

All of the reported experiments were subject to non-specific factors such as level of arousal, sensory sensitization, or motivation (hunger or thirst). The design of all of them also provided for measurement of only two options, giving the appearance of learning a 50% probability by chance on any trial. In none of the reported experiments were controls sufficient to rule out non-specific behavioral tendencies. Despite these weaknesses, the fact that these marginal results were accepted for publication by journals of the stature of *Nature, Science,* and *Proceedings of the National Academy of Sciences* reflects the seriousness with which the gatekeepers of scientific credibility took these reports. That, in turn, I would speculate, was attributable to three factors: (1) a propensity at the time to generalize the informational role of macromolecules from the genetic to the neural dimension, absent evidence to the contrary; (2) ignorance of the nature of cognitive representation of information in the brain, so that information storage in a macromolecular code could not then, and still cannot, be ruled out; and (3) the attribution to brain extracts of specific information rather than modulatory influences was a more dramatic and exciting interpretation of the results. The overarching fact that made thinking of this type easier was the spectacular, recent success of molecular biology in unraveling the secret of information storage in the gene, a problem seemingly no less intractable than the mechanism of memory.

While the experiments of Fjerdingstad, Reinis, and Jacobson appear to have been inspired by McConnell's work with planaria and the general influence of Hydén's evidence for nucleic acid changes related to brain plasticity, Ungar approached the problem from a different direction. Trained in classic pharmacology in the European tradition, he had been seeking evidence for an endogenous compound that antagonized the analgesic properties of opiates, thereby explaining, perhaps, the development of drug tolerance. Upon finding what he thought was evidence of such a substance using the guinea pig ileum bioassay to monitor opiate action (Cohen et al., 1965; Ungar & Cohen, 1966), it had occurred to him that behavioral extinction, or loss of response to a repeated stimulus, was analogous to the loss of drug sensitivity, and therefore subject to experimentation in the same way. This was the rationale behind his experiment to test for the presence of a substance in the brain that could lead to loss of the startle reflex (habituation) to a sudden loud sound (Ungar & Oceguera-Navarro, 1965).

Thus, Ungar was predisposed by a lifetime of research on proteins and peptides to look to them as repositories of information (Ungar, 1963), but the scientific community overall showed greater fixation on the nucleic acids. So while Ungar's experiment was relatively ignored, many labs took up immediately the attempt to replicate the experiments with RNA. Some labs followed the protocols reported by Fjerdingstad, Jacobson, and their colleagues closely (Byrne et al., 1966; Gordon et al., 1966; Gross & Carey, 1965; Halas et al., 1966), and some did not (Byrne et al., 1966; Luttges et al., 1966). Some succeeded in replicating the positive results to varying degrees, but many did not (Byrne et al., 1966; Luttges et al., 1966; Quarton, 1967). A flood of experimental results, both positive and negative, were published over the remainder of the 1960s (Chapouthier, 1983; Ungar, 1973; Ungar & Irwin, 1968). Soon it was recognized, even by the authors of the original studies (Nissen et al., 1965; Ungar & Irwin, 1967), that the transfer phenomenon, to the extent that it was real, was a complicated phenomenon (Adám & Faiszt, 1967; Carran & Nutter, 1966; Chapouthier & Ungerer, 1968; Kleban et al., 1968).

One of the earliest and most prolific students of the transfer experiments was Frank Rosenblatt, already well known as the creator of the Perceptron, the first computerized neural network capable of learning (Rosenblatt, 1958). Rosenblatt was a brilliant experimentalist with a critical but open mind and a penchant for complicated experimental designs. Like Ungar, he felt the evidence pointed much more strongly toward peptides than nucleic acids as transfer agents (Rosenblatt, Farrow & Herblin, 1966a), and his experiments attempted to methodically dissect the critical factors in achieving successful transfer of behavioral information with peptides (Rosenblatt, 1969; Rosenblatt, Farrow & Rhine, 1966b; Rosenblatt, Farrow & Rhine, 1966c; Rosenblatt & Miller, 1966d; Rosenblatt & Miller, 1966e). His involvement in the transfer field was significant, as the first computer scientist steeped in the tradition of connectionism who saw the potential of biochemical coding superimposed on hard-wired circuitry. Tragically, he died in a boating accident in 1971, before the results of his work had completely unfolded and his perceptive insights fully matured.

The publication of negative resuls by so many respected scientists put a damper on the excitement over the transfer experiments. But while enthusiasm for them waned within the scientific community generally and in the public mind, several workers continued to pursue the nuances of the experiments, in hopes of unraveling the critical variables that would better explain both the negative and positive results. This included the work of Arnold Golub (Golub, 1972; Golub et al., 1970), who had begun his experiments with Jim Dyal at Texas Christian University (Dyal, Golub & Marrone, 1967). Also, paradoxically, despite the widespread perception that the transfer results were a marginal and erratic phenomenon at best, federal grant support for the work continued at a relatively high level (Stern, 1999). In 1967, grants totaling over $150,000, amounting to 25% of all NIH funding for research on biochemical aspects of memory, were awarded for experiments using the transfer paradigm. The NIH awarded an additional $240,000 in grants from 1968 through 1971. Ungar alone received $359,482 from government agencies from 1967 through 1973. This amounted to 40% of the total funds dispensed for memory-transfer research (Stern, 2005). By this measure of peer evaluation, behavioral transfer was regarded at the time as a scientific finding of considerable potential.

As controversy had swirled around the transfer experiments, Ungar had sought to counter the impression that they represented a radical departure from established views of how the brain works. His basic idea relied on the growing conviction at the time that chemical information must mediate the highly specific nature of synaptic connections, as established in the retinotectal and sensoricortical pathways, for instance (Roberts & Flexner, 1966; Sperry, 1958). The review chapter that he and I coauthored (Ungar & Irwin, 1968) relied heavily on those findings, and the logic was extended to central synapses mediating behavior most lucidly by Ungar (1968b) in an essay in *Perspectives in Biology and Medicine.* He accepted the notion of a Hebbian synapse, and viewed the transfer molecules essentially as the manifestation of the "metabolic change . . . such that A's efficiency, as one of the cells firing B, is increased" (Hebb, 1949).

As debate over the first wave of transfer experiments raged, a clever strategy for studying changes in nucleic acid base composition was introduced by John Wilson, a biochemist who teamed up with Edward Glassman, a psychologist at the University of North Carolina. Wilson's strategy was to use double labels for nucleic acid precursors taken up during training, whereby one radioisotope (say ^{3}H) would be used in controls, while another (say ^{14}C) would be used in experimentals (Zemp et al., 1966). Brain tissue from both controls and experimentals would be processed together following training, mitigating any differential procedural effects. Initial results appeared promising (Adair, Wilson & Glassman, 1968; Zemp, Wilson & Glassman, 1967), and the work at North Carolina was highly regarded, since it appeared to combine thoughtful experimental design without challenging fundamental assumptions about neurochemical substrates of behavior. A review of the biochemical basis of learning and memory by Glassman (1969) was regarded as the definitive statement on the subject at the time. As with the transfer work, however, when details of the experiments were probed by an expanded group of investigators, the phenomena appeared less consistent and more complex (Dunn, 1976b; Entingh et al., 1974; Uphouse, MacInnes & Schlesinger, 1974). While this growing skepticism cast doubt over the interpretation of the original results, the apparent reality of changes in RNA turnover with learning was verified (Smith, 1975; Smith, Heistad & Thompson, 1975), and a more sophisticated appreciation of the likely role of other metabolites and pathways emerged, particularly for the glycolipids and glycoproteins that incorporate some of the same precursors as the nucleic acids (Dunn et al., 1974).

In 1967, a method for conditioning rats to avoid their natural tendency to enter a dark compartment through negative (electric shock) reinforcement was published by Gay & Raphelson (1967). It initially attracted little attention, but when Ungar read about it and found in his lab that this form of learning was extremely robust and reproducible (Ungar, Galvan & Clark, 1968), he decided to adopt the paradigm as his vehicle for isolating, purifying, and sequencing the peptide or peptides that he thought would be found to be repositories of information that would affect the behavior of naïve recipients. His ultimate goal was to identify a chemical substance that had a clear-cut biological effect (Ungar, 1970). His lack of attention to the nuances of behavioral research derived not so much from his lack of experience in experimental psychology as from his extensive experience in pharmacology and endocrinology, where bioassays provided guideposts at every stage of the process for isolating and characterizing an efficacious compound, with the compound itself being the endpoint, not the guideposts along the way (Ungar, 1973).

As work began in earnest to isolate and purify the transfer factor for dark avoidance, Ungar enlisted the help of Dominic Desiderio, a talented mass spectrometrist who had provided the critical expertise for solving the enigmatic structure of the thyroid hormone releasing factor with Roger Guillemin (Burgus et al., 1969; Burgus et al., 1970). A synthetic organic chemist at the University of Houston, Wolfgang Parr, was recruited as a third member of the team. After several preliminary reports hinting at partial success, a full report of a definitive sequence for "scotophobin," the peptide alleged to induce dark avoidance was published in *Nature* on 28 July 1972 (Ungar, Desiderio & Parr, 1972a). However, the manuscript had been submitted a full 18 months earlier. The long delay had been due to an unusual process in which the reviewer of the paper, Walter W. Stewart, had engaged in a lengthy correspondence with Ungar in an effort to resolve aspects of the research that Stewart considered questionable. At the time, Stewart was a staff scientist at the National Institutes of Health, without a doctoral degree, with few publications, and no evident experience in physiological psychology, but with unquestioned expertise in the mass spectrometric methods upon which elucidation of the structure of scotophobin depended. In an even more unusual practice, the editor of *Nature* decided to publish the paper as submitted by Ungar, Desiderio, and Parr, but to follow it immediately with a long critique by Stewart (1972), followed by a brief rebuttal from Ungar, Desiderio & Parr (1972b). Though Ungar thought the unusual publishing practice had worked to his advantage (Ungar, 1977), the fact is that his NIH grant was not renewed in 1972, and scotophobin was not generally accepted as a neuroactive peptide by the scientific community at large.

There was, in fact, a multitude of publications either confirming or denying the efficacy of scotophobin (Garfield, 1975; Stern, 1999). In addition, other molecules were reported to transfer different behavioral properties, consistent with Ungar's view that different behaviors must be "coded for" by different peptides (Table 1). These included molecules transferring a tendency by goldfish to respond to either blue or green stimuli (Tate, Galvan, and Ungar, 1976; Ungar, Galvan and Chapouthier, 1972c; Zippel and Domagck, 1969), and the factor (ameletine) responsible for the original report by Ungar and Oceguera-Navarro (1965) of habituation to sound (Burzynski, 1976).

Table 1. Amino acid sequences reported for behaviorally active peptides

Scotophobin	N-Ser-Asp-Asn-Asn-Gln-Gln-Gly-Lys- -Ser-Ala-Gln-Gln-Gly-Gly-TyrNH$_2$	Ungar et al. (1972)
Scotophobin	N-Ser-Glu-Gly-Lys- -Ser-Ala-Gln-Gln-Gly-Gly-TyrNH$_2$	Ungar (1976)
Chromodiopsin Green→Blue	N-AcLys-Gly-Glu-Ile-Ala-Val-Phe-Pro- -Leu-Lys-Tyr-Gly-SerOH	Tate, Galvan, Ungar (1976)
Chromodiopsin Blue→Green	N-pGlu-Ile-Gly--------Ala-Val-Phe-Pro- -Leu-Lys-Tyr-Gly- Ser-LysOH	Tate, Galvan, Ungar (1976)
Ameletine	N-*p*Glu-Ala-Gly-Tyr-Ser-LysOH	Burzynski (1976)

Except for these limited initiatives, however, the apparent promise of discovering a family of neuroactive peptides capable of modifying specific behavioral tendencies was gradually abandoned. Even Ungar appears to have recognized the futility of the effort once he lost his grant and was forced to give up his lab at the Baylor College of Medicine. When William Byrne generously provided space for him to continue at the University of Tennessee Health Science Center, he devoted the few remaining months of his life to searching for endogenous anti-opiates in the brain (Ungar & Ungar, 1976; Ungar et al., 1977).

Although no instance of transfer of specific behavioral information by injection of either nucleic acids or peptides into naïve animals has stood the test of time, the original implication of the neuroactive properties of peptides has become a staple of orthodoxy in neuroscience. That trend had begun as long ago as the first transfer experiments, with the pioneering work of Bela Bohus and David De Wied on the role of adrenocorticotropic hormone (ACTH) and its analogues in facilitating learning (Bohus et al., 1968; De Wied, 1965). A long series of studies gradually strengthened the evidence for the involvement of a variety of neuropeptides in the learning process (Bohus, Gispen & De Wied, 1973; De Wied, 1977; De Wied et al., 1978), including especially the work of Wilhelm Gispen (Bohus et al., 1973; Gispen et al., 1976) and Adrian Dunn (Dunn and Gispen, 1977; Dunn, Iuvone & Rees, 1976). In addition to ACTH and its analogues, however, many other peptides were eventually found to be neuroactive (Bloom et al., 1976; Chance et al., 1998; De Wied, 1997; De Wied & Gispen, 1977; Kovacs & De Wied, 1997; Morgan & Routtenberg, 1977; Thiele et al., 1997; Yamamoto et al., 1998). This succession of discoveries encountered little controversy, since they were viewed from the outset as serving a modulatory role rather than as coding molecules for specific behaviors or pathways. It is interesting to speculate whether the transfer experiments would have been received more sympathetically, had the mindset of the times not been so focused by the success of molecular biology on "informational molecules," and had Ungar not persisted in viewing the transfer factors in that way himself. A modulatory role for peptides would have been a reasonably conventional explanation for the transfer phenomenon that in retrospect may have led to a much more fruitful line of research. Indeed, that is essentially the interpretation eventually reached by several who continued to try to explain what really happens when peptide extracts from brains of trained animals are injected into naïve recipients (Chapouthier, 1983; Malin, 1974; Malin, Radcliffe & Osterman, 1976; Misslin et al., 1978; Wojcik & Niemierko, 1978).

The largest factor in the demise of the transfer approach, however, probably came from a fundamentally similar methodology that led to a dramatically more successful outcome—the discovery of endogenous opiates in the brain (Snyder & Childers, 1979). Like Ungar's approach to the search for a factor responsible for drug tolerance, research that led to the discovery of endogenous opiates had its roots in traditional pharmacology. Avram Goldstein was a formative figure in establishing the principle of drug receptor specificity, and pointed the way to an experimental method for demonstrating it (Goldstein, 1973). By the early 1970s, it was clear that opiate receptors must exist in specific areas of the brain. Using a version of the technique described earlier by Goldstein, but modified by using drugs with high specific activities, Candace Pert and Solomon Snyder demonstrated the presence of a specific receptor in the brain (Pert & Snyder, 1973). Almost simultaneously, Lars Terenius (Appelgren & Terenius, 1973; Terenius, 1973) and Eric Simon (Simon, Hiller & Edelman, 1973) provided similar evidence.

Even as the existence of specific opiate receptors was being confirmed, the search for an endogenous ligand to the receptors was underway (Terenius & Wahlstrom, 1975). Winning this race was the team of Hans Kosterlitz (1977) in Aberdeen, Scotland. The structures of the first two endogenous opiates, met-enkephalin and leu-enkephalin, were reported by Hughes *et al.* (1975). Other endogenous opiates, subsequently referred to generically as endorphins, were reported (Bloom et al., 1976; Cox et al., 1976a; Cox, Goldstein & Li, 1976b; Goldstein & Cox, 1977; Guillemin et al., 1977), many of which derive from the parent compound, pro-opiomelanocortin. This molecule is precursor to many neuroactive peptides. These and peptides from other precursors are now known to modulate the performance of animals in training situations similar to those used in the transfer experiments (Baizman et al., 1979).

Attempts to replicate the transfer experiments were extensive and widely reported, apparently without great difficulty, for more than a decade. Throughout the late 1960s, grant support for transfer studies was strong as well (Stern, 1999). The promise of the transfer experiments was clearly taken seriously by much of the neuroscientific community, including by those such as Agranoff and Bennett who were skeptical from the beginning. Most scientists accorded the transfer work a fair and respectful hearing. Only a few publicly expressed disdain for the effort. David Hubel (1979), a nobel laureate for his work on the neurophysiology of visual information processing, provided a rare, discordant example in a sarcastic evaluation in 1979:

> A few years ago the notion was advanced that memories might be recorded in the form of large molecules . . . Few people familiar with the highly patterned specificity of connections in the brain took the idea seriously . . . The fad has died out, but the fact is that neurobiology has not always advanced or even stood still; sometimes there is momentary backsliding.

No fully satisfactory explanation for the effect of scotophobin, if and when it does accentuate dark avoidance, has been forthcoming, but intriguing suggestions have been offered. Wilson (1986) noted that the last three amino acids at the aminated C-terminal of scotophobin—glycine-glycine-phenylalanine—correspond to the first three amino acids at the N-terminal of enkephalin in reverse order—phenylalanine-glycine-glycine. While an interesting observation, the functional significance of this coincidence is unclear. Malin and his colleagues showed that scotophobin interacts with exposure to a dark chamber to promote a robust adrenocorticoid stress response (1976), and Satake and co-workers demonstrated that a synthetic scotophobin analogue promotes dark avoidance in goldfish if the pineal gland is intact, apparently because the drug inhibits an enzyme in the biosynthetic pathway for melatonin, causing an elevation of pineal N-acetylserotonin (Satake & Morton, 1979a; Satake & Morton, 1979b; Satake & Morton, 1979c). Neither clue strengthens the case for memory storage in the form of a single, discreet molecule.

In my view, the evidence that RNA ever had convincing memory transfer properties is very weak, but the intriguing possibility was raised by Smalheiser, Manev, & Costa (2001) that RNA injections may have induced destruction of normal mRNA through the small double-stranded RNA inhibition (RNAi) process now known to protect eukaryotic cells from viral infection. Such an RNAi process, which was totally unknown at the time, may have knocked down the expression of genes that did in fact affect the function of brain cells. Thus,

while the explanation offered originally remains unconvincing, the phenomenology may have had a basis in fact.

Because this book has focused narrowly on the strategy of biochemical
transfer of behavioral information, much excellent work on the neural basis of
learning and memory during the period has not been mentioned. For the sake of
perspective, a few of the more notable examples will be referenced briefly.

William Greenough, following his doctoral work at Berkeley, pursued the
strategy of environmentally stimulating animals as a means of inducing neural
plasticity. He established beyond a doubt what his mentors had first suggested:
that microanatomical changes occur when animals process a richer repertoire of
information (Greenough, Hwang & Gorman, 1985; Ivanco & Greenough, 2000;
Volkmar & Greenough, 1972; Wallace et al., 1992). His student, Kristin Harris
(Harris, 1995; Harris, 1999; Harris et al., 1980), in turn extended this work and
became a leader in the field of microanatomical plasticity.

Much of the work through the 1970s and 1980s was devoted to a correla-
tional strategy, in which biochemical signatures of plasticity were sought without metabolic intervention. My own work proceeded along these lines, first on
my own (Irwin, 1969; Irwin & Samson, 1971), then with Robin Barraco at the
Wayne State University School of Medicine (Barraco & Irwin, 1976; Irwin,
Barraco & Terrian, 1978; Irwin & Terrian, 1978), where we focused especially
on glycoproteins because of the clear demonstration of their importance in cell-
cell recognition dating back to the work of Barondes (1976). Enjoying greater
success in the same effort was Hansjürgen Matthies and his colleagues, whose
work provided the strongest evidence for changes in glycoprotein structures
correlated with brain plasticity (Jork, Lossner & Matthies, 1978; Matthies, 1989;
Pohle et al., 1979; Popov et al., 1980). More recent evidence of the role of
glycosylated cell adhesion molecules in synaptic plasticity (Knafo et al., 2005;
Murai, Misner & Ranscht, 2002; Schmidt, 1995) lends support to these earlier
efforts.

The laboratory of Steven P. R. Rose at the Open University in Great Britain
has been very productive in the study of one-trial learning in the chick (Bateson,
Horn & Rose, 1969; Bateson, Horn & Rose, 1972; Bateson, Rose & Horn,
1973), and his efforts continue to the present time to beneficially exploit this
model (Anokhin, Tiunova & Rose, 2002; Johnston & Rose, 2001; Salinska,
Bourne & Rose, 2004). A scholar of both depth and breadth, Rose was
unsympathetic to the potential of the memory transfer experiments. Otherwise,
he has consistently been one of the most insightful thinkers about the
overwhelmingly difficult challenge of mapping events at the molecular level
onto phenomenology in the cognitive realm (Rose, 1981; Rose, 1991).

Victor Shashoua at the McLean Hospital in Massachusetts used a form of
learning reminiscent of Hydén's wire balancing task, in which fish were forced
to learn the complex motor skill of overcoming the disorienting effect of a float
attached to their ventral surface. This task was first used to study changes in
nucleic acids (Shashoua, 1968; Shashoua, 1970), without very credible results,

but subsequent study of proteins revealed an increased synthesis in the ependy-
mal cells of the brain ventricles of several factors designated as "ependymins"
(Shashoua, 1976; Shashoua, 1978; Shashoua, 1979). These appear to function as
cell adhesion molecules, and their study has continued, most notably in the
hands of Rupert Schmidt (Konigstorfer et al., 1989; Schmidt, 1995; 1995).

Probably the greatest insights into the neurobiology of learning in vertebrates has come from the discovery of long-term potentiation (LTP), initially in
the hippocampus (Alger & Teyler, 1976; Andersen et al., 1977; Bliss & Lomo,
1973) but later in other cortical regions as well (Bear & Kirkwood, 1993;
Escobar, Alcocer & Chao, 1998; Lee, Weisskopf & Ebner, 1991), and long-term
depression (LTD) in the cerebellum (Ito, 1986; Ito, 1989). By the 1970s, the
hippocampal slice preparation, first developed in Norway (Andersen, Bliss &
Skrede, 1971; Andersen et al., 1969) and brought by Timothy Teyler to the
United States (1976; Teyler, 1980), had become established as a preparation
appropriate for the study of neural plasticity (Baudry et al., 1981; Browning et
al., 1979; Lee et al., 1979; Lynch, Halpain & Baudry, 1982; Spencer et al.,
1976). An appropriate slice through the hippocampus provides access to a region
of the brain with well-defined multisynaptic pathways, where the concept of the
Hebbian synapse can be tested in specific neuronal populations (Lynch, 2003).
Both LTP and LTD seemed to fit the criteria of the Hebbian synapse (Thompson
et al., 1998), and a highly productive line of neurochemical and neuropharma-
cological dissection of those examples of plasticity ensued. Similar progress was
made by Richard Thompson and his colleagues using the eye-blink reflex to elucidate mechanisms of plasticity in the limbic system, cerebellum, and brain stem
(Berger, Alger & Thompson, 1976; McCormick et al., 1982; Thompson, 1991).

McConnell's instinct to study the simplest system of learning he could find
in the animal kingdom eventually took form, not in planaria, but in mollusks
with nervous systems hardly more complex than those of flatworms. Eric
Kandel and many colleagues succeeded in identifying specific neurons that mediate elementary forms of plasticity in the sea hare, *Aplysia californica* (Abrams,
Karl & Kandel, 1991; Carew, Castellucci & Kandel, 1971; Carew, Walters &
Kandel, 1981; Kandel et al., 1983; Walters, Carew & Kandel, 1979). Daniel
Alkon and others had similar success with *Hermissenda* (Crow & Alkon, 1978;
Neary, Crow & Alkon, 1981; Nelson & Alkon, 1988; Nelson, Collin & Alkon,
1990). These breakthroughs eventually led to a reasonably coherent explanation
of simple forms of learning from the molecular to the behavioral level. This is
the only work in over a century of speculation and research on the mechanisms
of learning and memory that has led to a Nobel Prize, awarded to Eric Kandel in
2000. It is perhaps a commentary on the intractability of the problem of memory
that the only Nobel Prize for work in the field has honored research on the
simplest form of learning, in organisms that have what can only charitably be
called a brain.

Fred Samson had nothing to do with research on any aspect of learning and
memory, but everything to do with my experience of it, as detailed in the personal odyssey described earlier. He was typical of most neuroscientists during

the 1960-1980 era, aware of the transfer work, and often opinionated about it, but not participating himself. Trained as a cell physiologist at the University of Chicago (Samson, Katz & Harris, 1955), he spent his entire academic career at the University of Kansas. His first major discovery was the degree to which newborn rats are capable of surviving anoxia (Samson & Dahl, 1957). This led to a productive line of research over two decades on energy metabolism in the brain. With his long-time colleagues, William Balfour and Nancy and Dennis Dahl (Dahl & Samson, 1959; Lolley, Balfour & Samson, 1961; Lolley & Samson, 1962; Samson, Dick & Balfour, 1964; Samson, Balfour & Dahl, 1959a; Samson, Balfour & Dahl, 1960a; Samson, Balfour & Jacobs, 1960b), he did more than anyone to elucidate the central role of adenosine triphosphatase in managing the energy budget of brain cells (Hexum, Samson & Himes, 1970; Samson, 1965; Samson & Quinn, 1967). The confrontation that left him with a permanent skepticism toward Georges Ungar occurred in 1957, in the course of a project with Harold Himwich on biochemical responses of the brain to hypoglycemia (Samson et al., 1959b). Much but not all the material on Samson in this book appeared in a previous biographical sketch (Irwin, 1992).

The only paper Samson and I published together was an outgrowth of my doctoral dissertation on the effect of behavioral stimulation on brain ganglioside turnover (Irwin & Samson, 1971). He supported my interest in behavioral neurochemistry unfailingly, but kept his own work focused at a more cellular level. Following his introduction to the role of tubulin in axoplasmic transport at a workshop chaired by Samuel Barondes at the Neurosciences Research Program (Barondes, 1967), he embarked on a new line of research, contributing a great deal to the biochemistry of tubulin (Samson, 1976; Samson, 1971), and stimulating many students (Fernandez, Burton & Samson, 1971; Hinkley & Samson, 1974; Redburn, Poisner & Samson, 1972; Twomey & Samson, 1972) and colleagues (Himes et al., 1976) to do likewise. He moved to the University of Kansas Medical School in Kansas City in 1972, and pursued new lines of research on the neurotoxicology of acetylcholinesterase inhibitors (Nelson et al., 1978), regional glucose metabolism (Pazdernik et al., 1985), and osmoregulation in the brain (Wade et al., 1988). Following his retirement in 1990, he continued the "conceptual research" that he had perfected with Francis O. Schmitt at the Neurosciences Research Program (NRP), focusing especially on the action of free radicals in biological systems (Pazdernik et al., 1992; Samson & Nelson, 2000).

Schmitt died in 1995, after transferring leadership of the NRP to Gerald Edelman, who moved it first to New York, then to San Diego. Whatever the ultimate judgment would be regarding the impact of the NRP on the development of neuroscience—and those judgments were varied—the catalytic role played by Schmitt cannot be denied, and the impact of the program after he was gone was much less than when he was its head and inspiration.

James V. McConnell remained at the University of Michigan for his entire career. After his research program diminished, he wrote a widely adopted textbook in Psychology that made him rich. He died in 1990.

David Krech had been the conceptual inspiration for the pioneers in the study of brain plasticity at Berkeley, so his sudden death in 1977 at the age of 68 left a significant void. But Rosenzweig and Bennett by then had taken up the problem of the role of protein synthesis in memory consolidation. Bennett and his colleagues (Flood et al., 1973; Flood et al., 1974) had discovered that aniso-mycin, a less toxic inhibitor of protein synthesis, could block long-term memory if administered prior to training, thus demonstrating decisively what Flexner, Agranoff, and Barondes had shown earlier but more tentatively (Mizumori et al., 1987; Mizumori, Rosenzweig & Bennett, 1985; Rosenzweig & Bennett, 1996). Marion C. Diamond continued to study neocortical plasticity, adding significant findings on the influence of hormones, sex, and neural-immune interactions on the structure of the brain (Diamond et al., 2001; Gaufo & Diamond, 1997; McShane et al., 1988).

Bernard Agranoff continued to be an influential figure in neurochemistry, serving as president of the American Society of Neurochemistry, and a co-editor of the field's most influential treatise, *Basic Neurochemistry* (Siegel, Agranoff & Albers, 1998). He branched into new areas of research on neural regeneration (Agranoff, 1977; Ballestero et al., 1999; Hieber, Agranoff & Goldman, 1992), while returning as well to his earliest roots in the biochemistry of lipids (Fisher, Heacock & Agranoff, 1992). He too spent his entire academic career at the University of Michigan.

After his early work on the protein requirement for memory consolidation, Samuel Barondes made important contributions to the field of axoplasmic tran-sport (Barondes, 1967; Barondes, 1968) and neuronal recognition, discovering a family of carbohydrate-binding molecules in slime mold with molecular rel-atives in brain cells (Barondes, 1970; Barondes & Rosen, 1976). He remained intent on relating brain research to psychiatry, and devoted much of his effort after moving to the University of California at San Francisco to writing books on that subject for a lay audience (Barondes, 1998; Barondes, 2005).

While their early studies on RNA and learning turned out to be equivocal, Edward Glassman and John Wilson subsequently pursued with greater success the study of functionally related phosphorylation of brain proteins (Perumal et al., 1977). Adrian Dunn studied various aspects of hormones, stress, and learn-ing for many years at the University of Florida (Dunn & Gispen, 1977), before moving to the Louisiana State University Health Science Center in Shreveport, where he enlarged his work to include the influence of cytokines on the brain (Ando & Dunn, 1999; Barkhudaryan & Dunn, 1999).

Nancy and Dennis Dahl were invaluable colleagues and close friends of Fred Samson, especially in the early years in Lawrence where as students they formed a highly productive team with their professor (Dahl, Dahl & Samson, 1956; Dahl & Samson, 1959; Dahl, Samson & Balfour, 1964; Samson et al., 1959a; Samson et al., 1960a; Samson et al., 1959b; Samson & Dahl, 1957). Nancy served on the faculty at the University of Kansas, inspiring two genera-tions of students, until her retirement in 1998. After battling cancer for years, she died in 2002, feisty to the end. When opportunities for research dried up for

Dennis, he completed a medical residency in Kansas City and returned to Lawrence, where he practiced medicine at the student health center for the remainder of his career.

After his years with Ungar, David Malin set up his own lab at the University of Houston—Clear Lake, and eventually showed that the endogenous neuropeptide FF may contribute to opiate tolerance (Lake et al., 1992). His work continued on neural mechanisms of tolerance and dependence for morphine (Malin et al., 1993), cocaine (Malin et al., 2000), and nicotine (Malin, 2001), including the role played by endogenous neuropeptides. Thus he ultimately realized what had been the essence of Georges Ungar's original goal in studying morphine tolerance.

In 1978, when memories of scotophobin were still fresh, I wrote an essay similar to this, which sought to honor the efforts of those who had built the research edifice of behavioral neurochemistry to that point, often through a sea of controversy, while acknowledging the shortcomings that inevitably attend the perspectives and limitations of knowledge at a particular time (Irwin, 1978). That essay was interpreted by some as a repudiation of the transfer work (Fjerdingstad, 1979). That was not my intent. Rather, it was a call to lower expectations to a more realistic level, using the power of new techniques within the limits of the questions they were capable of answering, but by no means abandoning our ultimate goal of understanding the physical basis of memory. I believe that eventually that understanding will come. Whether the behavioral transfer experiments will ultimately be seen as a misstep away from or a stride in the direction of realizing that goal will be a judgment of history. Let there not be a doubt, though, that I count it a privilege to have been part of the effort.

References

Abrams, T.W., Karl, K.A., Kandel, E.R. 1991. Biochemical studies of stimulus convergence during classical conditioning in Aplysia: dual regulation of adenylate cyclase by Ca^{2+}/calmodulin and transmitter. *J Neurosci* 11:2655–2665

Adair, L., Wilson, J., Glassman, E. 1968. Brain function and macromolecules, IV. Uridine incorporation into polysomes of mouse brain during different behavioral experiences. *Proc Natl Acad Sci USA* 61:917–922

Adám, G., Faiszt, J. 1967. Conditions for successful transfer effects. *Nature* 216:198–200

Adey, W.R. 1969. Slow electrical phenomena in the central nervous system. *Neurosci Res Program Bull* 7:79–83

Agranoff, B. 1965. Molecules and memories. *Perspect Biol Med* 9:13–22

Agranoff, B., Davis, R., Brink, J. 1965. Memory fixation in goldfish. *Proc Natl Acad Sci USA* 54:788–793

Agranoff, B., Davis, R., Brink, J. 1966. Chemical studies on memory fixation in goldfish. *Brain Res* 1:303–309

Agranoff, B., Davis, R., Casola, L., Lim, R. 1967. Actinomycin D blocks formation of memory of shock avoidance in goldfish. *Science* 158:1600–1

Agranoff, B., Klinger, P. 1964. Puromycin effect on memory fixation in the goldfish. *Science* 146:952–953

Agranoff, B.W. 1977. Approaches to the biochemistry of regeneration in the central nervous system. *Adv Exp Med Biol* 83:191–201

Alger, B., Teyler, T. 1976. Long-term and short-term plasticity in the CA1, CA3, and dentate regions of the rat hippocampal slice. *Brain Res* 110:463–480

Andersen, P., Bliss, T.V., Skrede, K.K. 1971. Lamellar organization of hippocampal excitatory pathways. *Exp Brain Res* 13:222–238

Andersen, P., Bliss, T.V., Lomo, T., Olsen, L.I., Skrede, K.K. 1969. Lamellar organization of hippocampal excitatory pathways. *Acta Physiol Scand* 76:4A–5A

Andersen, P., Sundberg, S.H., Sveen, O., Wigstrom, H. 1977. Specific long–lasting potentiation of synaptic transmission in hippocampal slices. *Nature* 266:736–737

Ando, T., Dunn, A.J. 1999. Mouse tumor necrosis factor–alpha increases brain tryptophan concentrations and norepinephrine metabolism while activating the HPA axis in mice. *Neuroimmunomodulation* 6:319–329

Anokhin, K.V., Tiunova, A.A., Rose, S.P.R. 2002. Reminder effects – reconsolidation or retrieval deficit? Pharmacological dissection with protein synthesis inhibitors following reminder for a passive–avoidance task in young chicks. *Euro J Neurosci* 15:1759–1765

Appelgren, L.E., Terenius, L. 1973. Differences in the autoradiographic localization of labelled morphine–like analgesics in the mouse. *Acta Physiol Scand* 88:175–182

Appleman, M., Kemp, R. 1966. Puromycin: a potent metabolic effect independent of protein synthesis. *Biochem Biophy Res Commun* 24:564–568

Babich, F.R, Jacobson, A.L, Bubash, S., Jacobson, A. 1965a. Transfer of a response to naive rats by injection of ribonucleic acid extracted from trained rats. *Science* 149:656–657

Babich, F., Jacobson, A., Bubash, S. 1965b. Cross–species transfer of learning: Effect of ribonucleic acid from hamsters on rat behavior. *Proc Natl Acad Sci USA* 54:1299–1302

Baizman, E.R., Cox, B.M., Osman, O.H., Goldstein, A. 1979. Experimental alterations of endorphin levels in rat pituitary. *Neuroendocrinology* 28:402–414

Ballestero, R.P., Dybowski, J.A., Levy, G., Agranoff, B.W., Uhler, M.D. 1999. Cloning and characterization of zRICH, a 2',3'–cyclic–nucleotide 3'–phosphodiesterase induced during zebrafish optic nerve regeneration. *J Neurochem* 72:1362–1371

Barkhudaryan, N., Dunn, A.J. 1999. Molecular mechanisms of actions of interleukin–6 on the brain, with special reference to serotonin and the hypothalamo–pituitary–adrenocortical axis. *Neurochem Res* 24:1169–1180

Barondes, S. 1965. Relationship of biological regulatory mechanisms to learning and memory. *Nature* 205:18–21

Barondes, S. 1967. Axoplasmic transport. *Neurosci Res Progr Bull* 5:307–364

Barondes, S. 1968. Incorporation of radioactive glucosamine into macromolecules of nerve endings. *J Neurochem.* 15:699–706

Barondes, S. 1970. Brain glycomacromolecules and interneuronal recognition. In: *The Neurosciences, Second Study Program.* F.O. Schmitt, editor. pp. 747–760. Rockefeller Univ. Press, New York

Barondes, S. 1998. Mood Genes: Hunting for Origins of Mania and Depression. W.H. Freeman, San Francisco

Barondes, S. 2005. *Better Than Prozac: Creating The Next Generation Of Psychiatric Drugs.* Oxford University Press, Oxford

Barondes, S., Cohen, H. 1966. Puromycin effect on successive phases of memory storage. *Science* 151:594–595

Barondes, S., Cohen, H. 1968b. Arousal and the conversion of "short-term" to "long-term" memory. *Proc Natl Acad Sci USA* 61:923–929

Barondes, S., Jarvik, M. 1964. The influence of actinomycin D on brain RNA synthesis and on memory. *J Neurochem* 11:187–195

Barondes, S., Rosen, S. 1976. Cell surface carbohydrate–binding proteins: role in cell recognition. *In:* Neuronal Recognition. S. Barondes, editor. pp. 331–356. Plenum Press, New York

Barraco, R.A., Irwin, L.N. 1976. Cerebral protein patterns from trained and naive pigeons. *J Neurochem* 27:393–397

Bateson, P., Horn, G., Rose, S.P.R. 1969. Effects of an imprinting procedure on regional incorporation of tritiated lysine into protein of chick brain. *Nature* 223:534–535

Bateson, P., Horn, G., Rose, S. P. R. 1972. Effects of early experience on regional incorporation of precursors into RNA and protein in the chick brain. *Brain Res* 39:449–465

Bateson, P., Rose, S. P. R., Horn, G. 1973. Imprinting: lasting effects on uracil incorporation into chick brain. *Science* 181:576–578

Baudry, M., Arst, D., Oliver, M., Lynch, G. 1981. Development of glutamate binding sites and their regulation by calcium in rat hippocampus. *Brain Res* 227:37–48

Bear, M.F., Kirkwood, A. 1993. Neocortical long-term poten–tiation. *Curr Opin Neurobiol* 3:197–202

Bennett, E.L., Calvin, M. 1964. Failure to train planarians reliably. *Neurosci Res Progr Bull* 2:3–24

Bennett, E.L., Diamond, M., Krech, D., Rosenzweig, M.R. 1964. Chemical and anatomical plasticity of brain. *Science* 146:610–619

Berger, T.W., Alger, B., Thompson, R.F. 1976. Neuronal substrate of classical conditioning in the hippocampus. *Science* 192:483–485

Bliss, T.V., Lomo, T. 1973. Long–lasting potentiation of synaptic transmission in the dentate area of the anaesthetized rabbit following stimulation of the perforant path. *J Physiol* 232:331–356

Bloom, F., Segal, D., Ling, N., Guillemin, R. 1976. Endorphins: profound behavioral effects in rats suggest new etiological factors in mental illness. *Science* 194:630–632

Bohus, B., Gispen, W.H., De Wied, D. 1973. Effect of lysine–vasopressin and ACTH 4–10 on conditioned avoidance behavior of hypophysectomized rats. *Neuroendocrinology* 11:137–143

Bohus, B., Nyakas, C., Endröczi, E. 1968. Effects of adrenocorticotropic hormone on avoidance behaviour of intact and adrenalectomized rats. *Int J Neuropharmacol* 7:307–314

Breen, R., McGaugh, J.L. 1961. Facilitation of maze learning with posttrial injections of picrotoxin. *J Compar Physiol Psychol* 54:498–501

Brenner, S., Jacob, F., Meselson, M. 1961. An unstable intermediate carrying information from genes to ribosomes for protein synthesis. *Nature* 190:576–581

Briggs, M. 1962. The molecular basis of memory and learning. *Psychol Rev* 69:537–541

Browning, M., Dunwiddie, T., Bennett, W., Gispen, W., Lynch, G. 1979. Synaptic phosphoproteins: specific changes after repetitive stimulation of the hippocampal slice. *Science* 203:60–62

Burgus, R., Dunn, T.F., Desiderio, D., Guillemin, R. 1969. [Molecular structure of the hypothalamic hypophysiotropic TRF factor of ovine origin: mass spectrometry demonstration of the PCA–His–Pro–NH2 sequence]. *C R Acad Sci Hebd Seances Acad Sci D* 269:1870–1873

Burgus, R., Dunn, T.F., Desiderio, D., Ward, D.N., Vale, W., Guillemin, R. 1970. Characterization of ovine hypothalamic hypophysiotropic TSH-releasing factor. *Nature* 226:321–325

Burzynski, S. 1976. Sequential analysis in subnanomolar amounts of peptides. Determination of the structure of a habituation–induced brain peptide (ameletine). *Analyt Biochem* 70:359–365

Byrne, W. L., Samuel, D., Bennett, E. L., Rosenzweig, M. R., Wasserman, E., Wagner, A. R., Gardner, R., Galambos, R., Berger, B. D., Margules, D. L., Fenichel, R. L., Stein, R., Corson, J. A., Enesco, H. E., Chorover, S. L., Holt, C. E., III, Schiller, P.H., Chiappeta, L., Jarvik, M., Leaf, R., Dutcher, J., Horovitz, Z., Carlson, P. 1966. Memory transfer. *Science* 153:658–659

Carew, T.J., Castellucci, V.F., Kandel, E.R. 1971. An analysis of dishabituation and sensitization of the gill–withdrawal reflex in Aplysia. *Int J Neurosci* 2:79–98

Carew, T.J., Walters, E.T., Kandel, E.R. 1981. Classical conditioning in a simple withdrawal reflex in Aplysia californica. *J Neurosci* 1:1426–1437

Carran, A., Nutter, C. 1966. Heredity–environment interaction in brain extract transfer in highly inbred mice. *Psychon Sci* 5:3–4

Chance, W.T., Tao, Z., Sheriff, S., Balasubramaniam, A. 1998. WRYamide, a NPY–based tripeptide that antagonizes feeding in rats. *Brain Res* 803:39–43

Chapouthier, G. 1983. Protein synthesis and memory. *In:* The Physiological Basis of Memory. J.A. Deutsch, ed., pp. 1–47. Academic Press, New York

Chapouthier, G., Ungerer, A. 1968. Effet de l'injection d'extraits de cerveau conditioné sur l'apprentissage. *Compt Rend Acad Sci Paris D* 269:769–771

Cohen, H., Barondes, S. 1967. Puromycin effect on memory may be due to occult seizures. *Science* 157:333–334

Cohen, M., Keats, A., Krivoy, W., Ungar, G. 1965. Effect of actinomycin D on morphine tolerance. *Proc Soc Exp Biol Med* 119:381–384

Corning, W. 1964. Evidence of a right–left discrimination in planarians. *J Psychol* 58:138–139

Corning, W. 1966. Retention of a position discrimination after regeneration in planarians. *Psychonom Sci* 5:17–18

Corning, W., John, E.R. 1961. Effect of ribonuclease on retention of conditioned response in regenerated planarians. *Science* 134:1363–1365

Cox, B.M., Gentleman, S., Su, T.P., Goldstein, A. 1976a. Further characterization of morphine–like peptides (endorphins) from pituitary. *Brain Res* 115:285–296

Cox, B.M., Goldstein, A., Li, C.H. 1976b. Opioid activity of a peptide, beta–lipotropin–(61–91), derived from beta–lipotropin. *Proc Natl Acad Sci USA* 73:1821–1823

Crow, T.J., Alkon, D.L. 1978. Retention of an associative behavioral change in Hermissenda. *Science* 201:1239–1241

Dahl, D.R., Dahl, N., Samson, F.E., Jr. 1956. A study on the narcotic action of the short chain fatty acids. *J Clin Invest* 35:1291–1298

Dahl, D.R., Samson, F.E., Jr. 1959. Metabolism of rat brain mitochondria during postnatal development. *Am J Physiol* 196:470–472

Dahl, N.A., Samson, F.E., Jr., Balfour, W.M. 1964. Adenosine Triphosphate and Electrical Activity in Chicken Vagus. *Am J Physiol* 206:818–822

De Wied, D. 1965. The influence of the posterior and intermediate lobe of the pituitary and pituitary peptides on the maintenance of the conditioned avoidance response in rats. *Int J Neuropharmacol* 4:157–167

De Wied, D. 1977. Peptides and behavior. *Life Sci* 20:195–204

De Wied, D. 1997. Neuropeptides in learning and memory processes. *Behav Brain Res* 83:83–90

De Wied, D., Bohus, B., Van Ree, J., Urban, I. 1978. Behavioral and electrophysiological effects of peptides related to lipotropin (β–LPH). *J Pharmacol Exp Therapeut* 204:570–580

De Wied, D., Gispen, W.H. 1977. Behavioral effects of peptides. *In:* Peptides in Neurobiology. H. Gainer, editor. pp. 397–. Plenum, New York

DeCamp, J. 1915. A study of retroactive inhibition. *Psychol Monogr* 19:1–69

Diamond, M.C., Krech, D., Rosenzweig, M.R. 1964. The effects of an enriched environment on the histology of the rat cerebral cortex. *J Compar Neurol* 123:111–120

Diamond, M.C., Weidner, J., Schow, P., Grell, S., Everett, M. 2001. Mental stimulation increases circulating CD4–positive T lymphocytes: a preliminary study. *Brain Res Cogn Brain Res* 12:329–31

Dingman, W., Sporn, M. 1961. The incorporation of 8–azaguanine into rat brain RNA and its effect on maze learning by the rat. *J Psychiat Res* 1:1–11

Dunn, A.J. 1976. The chemistry of learning and the formation of memory. *In.* Molecular and Functional Neurobiology. W.H. Gispen, editor. pp. 347–387. Elsevier, Amsterdam

Dunn, A.J., Entingh, D., Entingh, T., Gispen, W.H., Machlus, B., Perumal, R., Rees, H.D., Brogan, L. 1974. Biochemical correlates of brief behavioral experiences. In: *The Neurosciences: Third Study Program.* F.O. Schmitt and F.C. Worden, editors. MIT Press, Cambridge

Dunn, A.J., Gispen, W.H. 1977. How ACTH acts on brain. *Biobehav Rev* 1:15–23

Dunn, A.J., Iuvone, P.M., Rees, H.D. 1976. Neurochemical responses of mice to ACTH and lysine vasopressin. *Pharmac Biochem Behav* 5:139–145

Dyal, J., Golub, A., Marrone, R. 1967. Transfer effects of intraperitoneal injection of brain homogenate. *Nature* 214:720–721

Eccles, J.C. 1977. An instruction–selection theory of learning in the cerebellar cortex. *Brain Res* 127:327–352

Eccles, J.C. 1961. The mechanism of synaptic transmission. *Ergeb Physiol* 51:299–430

Eccles, J.C., Eccles, R.M., Fatt, P. 1956. Pharmacological investigations on a central synapse operated by acetylcholine. *J Physiol* 131:154–169

Eccles, J.C., McIntyre, A.K. 1953. The effects of disuse and of activity on mammalian spinal reflexes. *J Physiol* 121:495–516

Entingh, D., Damstra–Entingh, T., Dunn, A.J., Wilson, J., Glassman, E. 1974. Brain UMP: reduced incorporation of uridine during aboidance learning. *Brain Res* 70:131–138

Escobar, M.L., Alcocer, I., Chao, V. 1998. The NMDA receptor antagonist CPP impairs conditioned taste aversion and insular cortex long-term potentiation in vivo. *Brain Res* 812:246–251

Fernandez, H.L., Burton, P.R., Samson, F.E. 1971. Axoplasmic transport in the crayfish nerve cord. The role of fibrillar constituents of neurons. *J Cell Biol* 51:176–192

Fisher, S.K., Heacock, A.M., Agranoff, B.W. 1992. Inositol lipids and signal transduction in the nervous system: an update. *J Neurochem* 58:18–38

Fjerdingstad, E.J., Nissen, T., Roigaard–Petersen, H. 1965. Effect of ribonucleic acid (RNA) extracted from the brain of trained animals on learning in rats. *Scand J Psychol* 6:1–6

Fjerdingstad, E.J. 1979. The transfer paradigm. *Perspect Biol Med* 22:461–463

Flexner, J., Flexner, L., Stellar, E. 1963. Memory in mice as affected by intracerebral puromycin. *Science* 141:57–59

Flexner, J., Flexner, L., Stellar, E., de la Haba, G., Roberts, R. 1962. Inhibition of protein synthesis in brain and learning and memory following puromycin. *J Neurochem* 9:595–605

Flexner, L., Flexner, J. 1968. Studies on memory: The long survival of peptidyl–puromycin in mouse brain. *Proc Natl Acad Sci USA* 60:923–927

Flexner, L., Flexner, J., Roberts, R., de la Haba, G. 1964. Loss of recent memory in mice as related to regional inhibition of cerebral protein synthesis. *Proc Natl Acad Sci USA* 52:1165–1169

Flood, J.F., Rosenzweig, M.R., Bennett, E.L., Orme, A.E. 1973. The influence of duration of protein synthesis inhibition on memory. *Physiol Behav* 10:555–62

Flood, J.F., Rosenzweig, M.R., Bennett, E.L., Orme, A.E. 1974. Comparison of the effects of anisomycin on memory across six strains of mice. *Behav Biol* 10:147–160

Franklin, R.E., Gosling, R.G. 1953. Molecular configuration in sodium thymonucleate. *Nature* 171:740–741

Freud, S. 1900. The Interpretation of Dreams. *In:* The Basic Writings of Sigmund Freud. A.A. Brill, editor. pp. 487–491. Random House, New York

Gaito, J. 1963. DNA and RNA as memory molecules. *Psychol Rev* 70:471–480

Garfield, E. 1975. Using SCI to illuminate Scotophobin. *Curr Contents* 43:5–9

Gaufo, G.O., Diamond, M.C. 1997. Thymus graft reverses morphological deficits in dorsolateral frontal cortex of congenitally athymic nude mice. *Brain Res* 756:191–9

Gay, R., Raphelson, A. 1967. "Transfer of learning" by injection of brain RNA: a replication. *Psychon Sci* 8:369–370

Gispen, W., Buitelaar, J., Wiegant, V., Terenius, L., De Wied, D. 1976. Interaction between ACTH fragments, brain opiate receptors and morphine–induced analgesia. *Eur J Pharmacol* 39:393–397

Glassman, E. 1969. The biochemistry of learning: an evaluation of the role of RNA and protein. *Ann Rev Biochem* 38:605–646

Goldberg, A. 1964. Memory mechanisms. *Science* 144:1529

Goldstein, A. 1973. Interactions of narcotic antagonists with receptor sites. *Adv Biochem Psychopharmacol* 8:471–481

Goldstein, A., Cox, B. 1977. Opioid peptides (endorphins) in pituitary and brain. *Psychoneuroendocrinology* 2:11–16

Golub, A. 1972. Investigation of behavior induction by injection of mammalian brain extract. *In:* Methods in Neurochemistry. R. Fried, editor. Marcel Dekker, New York

Golub, A., Masiarz, F., Villars, T., McConnell, J. 1970. Incubation effects in behavior induction in rats. *Science* 168:392–395

Gordon, M., Deanin, G., Leonhardt, H., Gwynn, R. 1966. RNA and memory: a negative experiment. *Am J Psychiatry* 122:1174–1178

Greenough, W.T., Hwang, H.M., Gorman, C. 1985. Evidence for active synapse formation or altered postsynaptic metabolism in visual cortex of rats reared in complex environments. *Proc Natl Acad Sci USA* 82:4549–4552

Gros, F., Hiatt, H., Gilbert, W., Kurland, C., Risebrough, R., Watson, J. 1961. Unstable ribonucleic acid revealed by pulse labelling of *Escherichia coli*. *Nature* 190:581–585

Gross, C., Carey, F. 1965. Transfer of learned response by RNA injection: failure of attempts to replicate. *Science* 150:1749

Guillemin, R., Ling, N., Burgus, R., Bloom, F., Segal, D. 1977. Characterization of the endorphins, novel hypothalamic and neurohypophysial peptides with opiate–like activity: evidence that they induce profound behavioral changes. *Psychoneuroendocrinology* 2:59–62

Halas, E., Bradfield, K., Sandlie, M., Theye, F., Beardsley, J. 1966. Changes in rat behavior due to RNA injection. *Physiol Behav* 1:281–283

Halstead, W. 1951. Brain and intelligence. In: *Cerebral Mechanisms in Behavior*. L. Jeffries, editor. pp. 244–272. John Wiley, New York

Harris, K.M. 1995. How multiple–synapse boutons could preserve input specificity during an interneuronal spread of LTP. *Trends in Neurosciences* 18:365–369

Harris, K.M. 1999. Structure, development, and plasticity of dendritic spines. *Curr Opinion Neurobiol* 9:343–348

Harris, K.M., Cruce, W.L., Greenough, W.T., Teyler, T.J. 1980. A Golgi impregnation technique for thin brain slices maintained in vitro. *J Neurosci Methods* 2:363–371

Hebb, D.O. 1949. The Organization of Behavior: A Neuropsychological Theory. John Wiley, New York

Hechter, O. 1966. Reflections on the molecular basis of mental memory. *In:* Molecular Basis of Some Aspects of Mental Activity. O. Walaas, editor. pp. 7–27. Academic Press, New York

Hering, E. 1895. On Memory and the Specific Energies of the Nervous System. Open Court Publishing, Chicago

Hexum, T., Samson, F.E., Jr., Himes, R.H. 1970. Kinetic studies of membrane $(Na^+–K^+–Mg^{2+})$–ATPase. *Biochim Biophys Acta* 212:322–331

Hieber, V., Agranoff, B.W., Goldman, D. 1992. Target–dependent regulation of retinal nicotinic acetylcholine receptor and tubulin RNAs during optic nerve regeneration in goldfish. *J Neurochem* 58:1009–1015

Hilgard, E.R. 1956. Theories of Learning. Appleton–Century–Crofts, New York

Himes, R.H., Kersey, R.N., Heller–Bettinger, I., Samson, F.E. 1976. Action of the vinca alkaloids vincristine, vinblastine, and desacetyl vinblastine amide on microtubules in vitro. *Cancer Res* 36:3798–3802

Hinkley, R.E., Jr., Samson, F.E., Jr. 1974. The effects of an elevated temperature, colchicine, and vinblastine on axonal microtubules of the crayfish (*Procambarus clarkii*). *J Exp Zool* 188:321–36

Hodgkin, A.L., Huxley, A.F. 1952. A quantitative description of membrane current and its application to conduction and excitation in nerve. *J Physiol* 117:500–544

Hubel, D.H. 1979. The brain. *Sci Amer* 241:45–53

Hughes, J., Smith, T.W., Kosterlitz, H.W., Fothergill, L.A., Morgan, B.A., Morris, H.R. 1975. Identification of 2 related pentapeptides from brain with potent opiate agonist activity. *Nature* 258:577–579

Hydén, H. 1943. Protein metabolism in the nerve cell during growth and function. *Acta Physiol Scand* 6:1–136

Hydén, H. 1959. Biochemical changes in glial cells and nerve cells. *In:* Proceedings of the Fourth International Congress of Biochemistry. F. Brücke, editor. pp. 64–89. Pergamon, New York

Hydén, H. 1960. A cytophysiological study of the functional relationship between oligodendroglial cells and nerve cells of Dieter's nucleus. *J Neurochem* 6:57–72

Hydén, H. 1966. A genetic stimulation with production of adenic–uracil rich RNA in neurons and glia in learning. *Naturwissenschaften* 53:64–70

Hydén, H. 1967. Biochemical changes accompanying learning. *In:* The Neurosciences. G. Quarton, T. Melnechuk, and F.O. Schmitt, editors. pp. 765–771. Rockefeller University Press, New York

Hydén, H. 1970. The question of molecular basis for the memory trace. *In:* Biology of Memory. K. Pribram and D. Broadbent, editors. pp. 101–119. Academic Press, New York

Hydén, H. 1976. Plastic changes of neurons during acquisition of new behavior as a problem of protein differentiation. *Prog Brain Res* 45:83–100

Hydén, H., Egyházi, E. 1962. Nuclear RNA changes of nerve cells during a learning experiment in rats. *Proc Natl Acad Sci USA* 48:1366–1373

Hydén, H., Egyházi, E. 1963. Glial RNA changes of nerve cells during a learning experiment with rats. *Proc Natl Acad Sci USA* 49:618–624

Hydén, H., Egyházi, E. 1964. Change in RNA content and base composition in cortical neurons of rats in a learning experiment involving transfer of handedness. *Proc Natl Acad Sci USA* 52:1030–1035

Hydén, H., Egyházi, E., John, E.R. 1969. RNA base ratio changes in planaria during conditioning. *J Neurochem* 16:813–821

Hydén, H., Lange, P. 1965. A differentiation in RNA response in neurons early and late during learning. *Proc Natl Acad Sci USA* 53:946–952

Hydén, H., Lange, P., Perrin, C. 1977. Protein pattern alterations in hippocampal and cortical cells as a function of training in rats. *Brain Res* 119:427–437

Irwin, L.N. 1969. Biochemical changes in subcellular fractions of brains of stimulated rats. *Brain Res* 15:518–521

Irwin, L.N. 1978. Fulfillment and frustration: the confessions of a behavioral biochemist. *Perspect Biol Med* 21:476–491

Irwin, L.N. 1992. Acrobat to academic: a sketch of the life and career of Frederick E. Samson, Jr. *Neurochem Res* 17:5–10.

Irwin, L.N., Barraco, R. A., Terrian, D. 1978. Protein and glycoprotein metabolism in brains of operantly conditioned pigeons. *Neuroscience* 3:457–464

Irwin, L.N., Samson, F.E. 1971. Content and turnover of gangliosides in rat brain following behavioral stimulation. *J Neurochem* 18:203–211

Irwin, L.N., Terrian, D.M. 1978. Glucosamine incorporation into brain glycoproteins and gangliosides: effect of adrenalectomy, corticosterone, exercise, and training. *Pharmac Biochem Behav* 9:33–37

Ito, M. 1986. Long-term depression as a memory process in the cerebellum. *Neurosci Res* 3:531–539

Ito, M. 1989. Long-term depression. *Annu Rev Neurosci* 12:85–102

Ivanco, T.L., Greenough, W.T. 2000. Physiological consequences of morpho-logically detectable synaptic plasticity: potential uses for examining recovery following damage. *Neuropharmacology* 39:765–76

Jacob, F., Monod, J. 1961. Genetic regulatory mechanisms in the synthesis of proteins. *J Molec Biol* 3:318–356

Jacobson, A., Babich, F., Bubash, S., Goren, C. 1966a. Maze preferences in naive rats produced by injection of ribonucleic acid from trained rats. *Psychonom Sci* 4:3–4

Jacobson, A.L., Babich, F., Bubash, S., Jacobson, A. 1966b. Differential approach tendencies produced by injection of RNA from trained rats. *Science* 150:636–637

Jacobson, A., Fried, C., Horowitz, S. 1966c. Planarians and memory. *Nature* 209:599–601

John, E.R. 1961. Higher nervous functions: brain functions and learning. *Ann Rev Physiol* 23:451–484

John, E.R. 1972. Switchboard versus statistical theories of learning and memory. *Science* 177:850–864

Johnston, A.N.B., Rose, S.P.R. 2001. Memory consolidation in day–old chicks requires BDNF but not NGF or NT–3; an antisense study. *Molec Brain Res* 88:26–36

Jork, R., Lossner, B., Matthies, H. 1978. Hippocampal activation and incorporation of macromolecule precursors. *Pharmacol Biochem Behav* 9:709–712

Judson, H. 1979. The Eighth Day of Creation. Simon & Schuster, New York

Kandel, E.R., Abrams, T., Bernier, L., Carew, T.J., Hawkins, R.D., Schwartz, J.H. 1983. Classical conditioning and sensitization share aspects of the same molecular cascade in Aplysia. *Cold Spring Harb Symp Quant Biol* 48 Pt 2:821–830

Kleban, M., Altschuler, H., Lawton, M., Parris, J., Lorde, C. 1968. Influence of donor–recipient brain transfers on avoidance learning. *Psychol Repts* 23:51–56

Knafo, S., Barkai, E., Herrero, A.I., Libersat, F., Sandi, C., Venero, U. 2005. Olfactory learning–related NCAM expression is state, time, and location specific and is correlated with individual learning capabilities. *Hippocampus* 15:316–325

Konigstorfer, A., Sterrer, S., Eckerskorn, C., Lottspeich, F., Schmidt, R., Hoffmann, W. 1989. Molecular characterization of an ependymin precursor from goldfish brain. *J Neurochem* 52:310–312

Konorski, J. 1950. Mechanisms of learning. *Symp. Soc. Exp. Biol.* 4:409–431

Kosterlitz, H.W., Hughes, J. 1977. Peptides with morphine–like action in brain. *British J Psychiat* 130:298–304

Kovacs, G.L., De Wied, D. 1997. Learning, memory and neuropeptides. *Endocrinol Metabol* 4:97–98

Krech, D., Rosenzweig, M.R., Bennett, E.L. 1960. Effects of environmental complexity and training on brain chemistry. *J Compar Physiol Psychol* 53:509–519

Krech, D., Rosenzweig, M.R., Bennett, E.L., Krueckel, B. 1954. Enzyme concentrations in the brain and adjustive behavior patterns. *Science* 120:994–996

Lake, J.R., Hebert, K.M., Payza, K., Deshotel, K.D., Hausam, D.D., Witherspoon, W.E., Arcangeli, K.A., Malin, D.H. 1992. Analog of neuropeptide FF attenuates morphine tolerance. *Neurosci Lett* 146:203–206

Lashley, K. 1929. Brain Mechanisms and Intelligence: A Quantitative Study of Injuries to the Brain. University of Chicago Press, Chicago

Lashley, K. 1950. In search of the engram. *Symp Soc Exp Biol* 4: 454–482

Lee, K., Oliver, M., Schottler, F., Creager, R., Lynch, G. 1979. Ultrastructural effects of repetitive synaptic stimulation in the hippocampal slice preparation: a preliminary report. *Exp Neurol* 65:478–480

Lee, S.M., Weisskopf, M.G., Ebner, F.F. 1991. Horizontal long-term potentiation of responses in rat somatosensory cortex. *Brain Res* 544:303–310

Lolley, R.N., Balfour, W.M., Samson, F.E., Jr. 1961. The high–energy phosphates in developing brain. *J New Drugs* 7:289–297

Lolley, R.N., Samson, F.E., Jr. 1962. Cerebral high–energy compounds: changes in anoxia. *Am J Physiol* 202:77–79

Luttges, M., Johnson, T., Buck, C., Holland, J., McGaugh, J.L. 1966. An examination of "transfer of learning" by nucleic acid. *Science* 151:834–837

Lynch, G. 2003. Long-term potentiation in the Eocene. *Philos Trans Royal Soc Lond B Biol Sci* 358:625–628

Lynch, G., Halpain, S., Baudry, M. 1982. Effects of high–frequency synaptic stimulation on glumate receptor binding studied with a modified in vitro hippocampal slice preparation. *Brain Res* 244:101–111

Malin, D.H. 1974. Synthetic scotophobin: analysis of behavioral effects in mice. *Pharmac Biochem Behav* 2:147–153

Malin, D.H. 2001. Nicotine dependence: studies with a laboratory model. *Pharmac Biochem Behav* 70:551–9

Malin, D.H., Moon, W.D., Moy, E.T., Jennings, R.E., Moy, D.M., Warner, R.L., Wilson, O.B. 2000. A rodent model of cocaine abstinence syndrome. *Pharmac Biochem Behav* 66:323–8

Malin, D.H., Radcliffe, G.J., Osterman, D.M. 1976. Stimulus specific effect of scotophobin on mouse plasma corticoids. *Pharmacol Biochem Behav* 4:481–483

Malin, D.R., Zadina, J.E., Lake, J.R., Rogillio, R.B., Leyva, J.E., Benson, T.M., Corriere, L.S., Handunge, B.P., Kastin, A.J. 1993. Tyr–MIF–1 precipitates abstinence syndrome in morphine–dependent rats. *Brain Res* 610:169–171

Matthies, H. 1989. In search of cellular mechanisms of memory. *Progr Neurobiol* 32:277–349

McConnell, J. 1962. Memory transfer through cannibalism in planarians. *J Neuropsychiatry* 3:42–48

McConnell, J., Jacobson, A., Kimble, D. 1959. The effects of regeneration upon retention of a conditioned response in the planarian. *J Compar Physiol Psychol* 52:1–5

McCormick, D.A., Clark, G.A., Lavond, D.G., Thompson, R.F. 1982. Initial localization of the memory trace for a basic form of learning. *Proc Natl Acad Sci USA* 79:2731–2735

McGaugh, J.L. 1961. Facilitative and disruptive effects of strychnine sulphate on maze learning. *Psychol Repts* 8:99–104

McGaugh, J.L., Petrinovich, L. 1959. The effect of strychnine sulphate on maze–learning. *Amer J Psychol* 72:99–102

McGaugh, J.L., Thompson, C. 1962. Facilitation of simultaneous discrimination learning with strychnine sulphate. *Psychopharmacologia* 3:166–172

McGaugh, J.L., Westbrook, W., Thompson, C. 1962. Facilitation of maze learning with postrial injections of 5–7–diphenyl–1–3–diazadamantan–6–ol (1757 I.S.). *J Compar Physiol Psychol* 55:710–713

McShane, S., Glaser, L., Greer, E.R., Houtz, J., Tong, M.F., Diamond, M.C. 1988. Cortical asymmetry—a preliminary study: neurons–glia, female–male. *Exp Neurol* 99:353–361

Mihailovic, L., Jankovic, B.D., Petkovic, M., Isakovic, K. 1958. Effect of electroshock upon nucleic acid concentrations in various parts of cat brain. *Experientia* 14:144–145

Misslin, R., Ropartz, P., Ungerer, A., Mandel, P. 1978. Non–reproducibility of behavioral–effects induced by scotophobin. *Behavioural Processes* 3:45–56

Mizumori, S.J., Channon, V., Rosenzweig, M.R., Bennett, E.L. 1987. Anisomycin impairs long-term working memory in a delayed alternation task. *Behav Neural Biol* 47:1–6

Mizumori, S.J., Rosenzweig, M.R., Bennett, E.L. 1985. Long-term working memory in the rat: effects of hippocampally applied anisomycin. *Behav Neurosci* 99:220–232

Monné, L. 1948. Functioning of the cytoplasm. *Adv Enzymol* 8:1–65

Morgan, J., Routtenberg, A. 1977. Angiotensin injected into the neostriatum after learning disrupts retention performance. *Science* 196:87–89

Murai, K.K., Misner, D., Ranscht, B. 2002. Contactin supports synaptic plasticity associated with hippocampal long-term depression but not potentiation. *Curr Biol* 12:181–190

Neary, J.T., Crow, T., Alkon, D.L. 1981. Change in a specific phosphoprotein band following associative learning in *Hermissenda*. *Nature* 293:658–660

Nelson, P.G. 1967. Brain mechanisms and memory. *In:* The Neurosciences. G. Quarton, T. Melnechuk, and F.O. Schmitt, editors. pp. 772–775. Rockefeller University Press, New York

Nelson, S.R., Doull, J., Tockman, B.A., Cristiano, P.J., Samson, F.E. 1978. Regional brain metabolism changes induced by acetylcholinesterase inhibitors. *Brain Res* 157:186–190

Nelson, T.J., Alkon, D.L. 1988. Prolonged RNA changes in the *Hermissenda* eye induced by classical conditioning. *Proc Natl Acad Sci USA* 85:7800–4

Nelson, T.J., Collin, C., Alkon, D.L. 1990. Isolation of a G protein that is modified by learning and reduces potassium currents in *Hermissenda*. *Science* 247:1479–1483

Nirenberg, M.W., Matthaei, J.H., Jones, O.W., Martin, R.G., Barondes, S.H. 1963. Approximation of genetic code via cell–free protein synthesis directed by template RNA. *Fed Proc* 22:55–61

Nirenberg, M.W., Matthaei, J.H. 1961. The dependence of cell–free protein synthesis in *E. coli* upon naturally occurring or synthetic polyribonucleotides. *Proc Natl Acad Sci USA* 47:1588–1602

Nissen, T., Roigaard–Petersen, H., Fjerdingstad, E. 1965. Effect of ribonucleic acid (RNA) extracted from the brain of trained animals on learning in rats. II. Dependence of RNA effect on training conditions prior to RNA extraction. *Scand J Psychol* 6:265–272

Pavlov, I.V. 1927. Conditioned Reflexes. Dover, 1960, New York

Pazdernik, T.L., Cross, R.S., Giesler, M., Samson, F.E., Nelson, S.R. 1985. Changes in local cerebral glucose utilization induced by convulsants. *Neuroscience* 14:823–835

Pazdernik, T.L., Layton, M., Nelson, S.R., Samson, F.E. 1992. The osmotic/calcium stress theory of brain damage: are free radicals involved? *Neurochem Res* 17:11–21

Penfield, W. 1955. The twenty–ninth Maudsley lecture: the role of the temporal cortex in certain psychical phenomena. *J Ment Sci* 101:451–465

Perot, P., Penfield, W. 1960. Hallucinations of past experience and experiential responses to stimulation of temporal cortex. *Trans Am Neurol Assoc* 85:80–84

Pert, C.B., Snyder, S.H. 1973. Opiate receptor: demonstration in nervous tissue. *Science* 179:1011–1014

Perumal, R., Gispen, W.H., Glassman, E., Wilson, J.E. 1977. Phosphorylation of proteins of synaptosome–enriched fractions of brain during short-term training experience: biochemical characterization. *Behav Biol* 21:341–57

Pohle, W., Ruthrich, H., Popov, N., Matthies, H. 1979. Fucose incorporation into rat hippocampus structures after acquisition of a brightness discrimination. A histoautoradiographic analysis. *Acta Biol Med Ger* 38:53–56

Popov, N., Schulzeck, S., Pohle, W., Matthies, H. 1980. Changes in the incorporation of [3H]fucose into rat hippocampus after acquisition of a brightness discrimination reaction. An electrophoretic study. *Neuroscience* 5:161–167

Quarton, G. 1967. The enhancement of learning by drugs and the transfer of learning by macromolecules. *In:* The Neurosciences. G. Quarton, T. Melnechuk, and F.O. Schmitt, editors. pp. 744–755. Rockefeller University Press, New York

Ramón y Cajal, S. 1899. Histology of the Nervous System (Translated from the French version of the original in Spanish, Textura del Sistema Nervioso del Hombre y Vertebrados). Oxford Univ. Press (1995), New York

Redburn, D.A., Poisner, A.M., Samson, F.E., Jr. 1972. Comparison of microtubule protein (tubulin) from adrenal medulla and brain. *Brain Res* 44:615–624

Reinis, S. 1965. The formation of conditioned reflexes in rats after the parenteral administration of brain homogenate. *Activitas Nervosa Superior* 7:167–168

Roberts, R.A., Flexner, L. 1966. A model for the development of retina–cortex connections. *Amer Scientist* 54:174–183

Rose, S.P.R. 1981. What should a biochemistry of learning and memory be about? *Neuroscience* 6:811–821

Rose, S.P.R. 1991. How chicks make memories: The cellular cascade from c–fos to dendritic remodelling. *Trends Neurosci* 14:390–397

Rose, S.P.R. 1998. Brains, minds, and the world. *In:* From Brains to Consciousness? S.P.R. Rose, editor. pp. 1–17. Princeton University Press, Princeton

Rosenblatt, F. 1958. The Perceptron: A probabilistic model for information storage and organization in the brain. *Psychol Rev* 65:386–408

Rosenblatt, F. 1969. behavior induction by brain extracts: a comparison of two procedures. *Proc Natl Acad Sci USA* 64:661–668

Rosenblatt, F., Farrow, J., Herblin, W. 1966a. Transfer of conditioned responses from trained rats to untrained rats by means of a brain extract. *Nature* 209:46–48

Rosenblatt, F., Farrow, J., Rhine, S. 1966b. The transfer of learned behavior from trained to untrained rats by means of brain extracts, I. *Proc Natl Acad Sci USA* 55:548–555

Rosenblatt, F., Farrow, J., Rhine, S. 1966c. The transfer of learned behavior from trained to untrained rats by means of brain extracts, II. *Proc Natl Acad Sci USA* 55:787–792

Rosenblatt, F., Miller, R. 1966d. Behavioral assay procedures for transfer of learned behavior by brain extracts, Part I. *Proc Natl Acad Sci USA* 56:1423–1430

Rosenblatt, F., Miller, R. 1966e. Behavioral assay procedures for transfer of learned behavior by brain extracts, Part II. *Proc Natl Acad Sci USA* 56:1683–1688

Rosenzweig, M.R., Bennett, E.L. 1996. Psychobiology of plasticity: effects of training and experience on brain and behavior. *Behav Brain Res* 78:57–65

Rosenzweig, M.R., Bennett, E.L., Krech, D. 1964. Cerebral effects of environmental complexity and training among adult rats. *J Compar Physiol Psychol* 57:438–439

Rosenzweig, M.R., Krech, D., Bennett, E.L. 1958. Brain chemistry and adaptive behavior. *In:* Biological and Biochemical Bases of Behavior. H. Harlow and C. Woolsey, editors. pp. 367–400. University of Wisconsin Press, Madison

Rosenzweig, M.R., Krech, D., Bennett, E.L. 1960. A search for relations between brain chemistry and behavior. *Psychol Bull* 57:476–492

Rosenzweig, M.R., Krech, D., Bennett, E.L., Diamond, M.C. 1962. Effects of environmental complexity and training on brain chemistry and anatomy: a replication and extention. *J Compar Physiol Psychol* 55:429–437

Salinska, E., Bourne, R.C., Rose, S.P.R. 2004. Reminder effects: the molecular cascade following a reminder in young chicks does not recapitulate that following training on a passive avoidance task. *Europ J Neurosci* 19:3042–7

Samson, F.E. 1976. Pharmacology of drugs that affect intracellular movement. *Annu Rev Pharmacol Toxicol* 16:143–59

Samson, F.E., Dick, H.C., Balfour, W.M. 1964. Na^+–K^+ stimulated ATPase in brain during neonatal maturation. *Life Sci* 14:511–5

Samson, F.E., Jr. 1965. Energy flow in brain. *Prog Brain Res* 16:216–28

Samson, F.E., Jr. 1971. Mechanism of axoplasmic transport. *J Neurobiol* 2:347–360

Samson, F.E., Jr., Balfour, W.M., Dahl, N.A. 1959a. Cerebral free energy and viability: ATP in rats under nitrogen and iodoacetate with the effects of temperature. *Am J Physiol* 196:325–6

Samson, F.E., Jr., Balfour, W.M., Dahl, N.A. 1960a. Rate of cerebral ATP utilization in rats. *Am J Physiol* 198:213–216

Samson, F.E., Jr., Balfour, W.M., Jacobs, R.J. 1960b. Mitochondrial changes in developing rat brain. *Am J Physiol* 199:693–6

Samson, F.E., Jr., Dahl, D.R., Dahl, N., Himwich, H.E. 1959b. Studies of the hypoglycemic brain; amino acids, nucleic acids, total nitrogen, and side–group ionization of proteins in cat brain during insulin coma. *AMA Arch Neurol Psychiat* 81:458–65

Samson, F.E., Jr., Dahl, N.A. 1957. Cerebral energy requirement of neonatal rats. *Am J Physiol* 188:277–280

Samson, F.E., Jr., Quinn, D.J. 1967. Na^+–K^+–activated ATPase in rat brain development. *J Neurochem* 14:421–427

Samson, F.E., Katz, A.M., Harris, D.L. 1955. Effects of acetate and other short–chain fatty acids on yeast metabolism. *Arch Biochem* 54:406–423

Samson, F.E., Nelson, S.R. 2000. The aging brain, metals and oxygen free radicals. *Cell Mol Biol (Noisy–le–grand)* 46:699–707

Satake, N., Morton, B. 1979a. Pineal hydroxyindole–O–methyltransferase – mechanism, and inhibition by Scotophobin–A. *Pharmac Biochem Behav* 10:457–462

Satake, N., Morton, B.E. 1979b. Scotophobin–a causes dark avoidance in goldfish by elevating pineal N–acetylserotonin. *Pharmac Biochem Behav* 10:449–456

Satake, N., Morton, B.E. 1979c. Scotophobin–a causes several responses in goldfish if the pineal–gland is present. *Pharmacol Biochem Behav* 10:183–188

Schmidt, R. 1995. Cell–adhesion molecules in memory formation. *Behav Brain Res* 66:65–72

Schmidt, R., Brysch, W., Rother, S., Schlingensiepen, K.H. 1995. Inhibition of memory consolidation after active–avoidance conditioning by antisense intervention with ependymin gene–expression. *J Neurochem* 65:1465–1471

Shashoua, V. 1968. RNA changes in goldfish brain during learning. *Nature* 217:238–240

Shashoua, V. 1970. RNA metabolism in goldfish brain during acquisition of new behavioral patterns. *Proc Natl Acad Sci USA* 65:160–167

Shashoua, V. 1976. Brain metabolism and the acquisition of new behaviors. I. Evidence for specific changes in the pattern of protein synthesis. *Brain Res* 111:347–364

Shashoua, V. 1978. Effect of antisera to β and γ goldfish brain proteins on the retention of a newly acquired behavior. *Brain Res* 148:441–449

Shashoua, V. 1979. Brain metabolism and the acquisition of new behaviors. III. Evidence for secretion of two proteins into the brain extracellular fluid after training. *Brain Res* 166:349–358

Sherrington, C. 1906. The Integrative Action of the Nervous System. Yale University Press, New York

Siegel, G., Agranoff, B., Albers, R. 1998. Basic Neurochemistry: Molecular, Cellular and Medical Aspects. Lippincott Williams & Wilkins

Simon, E.J., Hiller, J.M., Edelman, I. 1973. Stereospecific binding of the potent narcotic analgesic (^{3}H)Etorphine to rat–brain homogenate. *Proc Natl Acad Sci USA* 70:1947–1949

Smalheiser, N., Manev, H., Costa, E. 2001. RNAi and brain function: was McConnell on the right track? *Trends Neurosci* 24:216–218

Smith, J. 1975. Distribution of ^{3}H–uridine–5 into brain RNA species of rats exposed to various training tasks – an electrophoretic analysis. *Pharmacol Biochem Behav* 3:455–461

Smith, J., Heistad, G., Thompson, T. 1975. Uptake of ^{3}H–uridine into brain and incorporation into brain RNA of rats exposed to various training tasks – a biochemical analysis. *Pharmac Biochem Behav* 3:447–454

Snyder, S., Childers, S. 1979. Opiate receptors and opioid peptides. *Ann Rev Neurosci* 2:35–64

Spencer, H.J., Gribkoff, V.K., Cotman, C.W., Lynch, G.S. 1976. GDEE antagonism of iontophoretic amino acid excitations in the intact hippocampus and in the hippocampal slice preparation. *Brain Res* 105:471–481

Sperry, R. 1958. Physiological plasticity and brain circuit theory. *In:* Biological and Biochemical Bases of Behavior. H. Harlow and C. Woolsey, editors. pp. 401–422. University of Wisconsin Press, Madison

Stern, L. 1999. Reception of "Extraordinary Scientific Claims": Georges Ungar, Scotophobin, and the Search for a Molecular Code for Memory. *In:* ISHPSSB Meetings, Oaxaca, Mexico

Stern, L. 2005. The Memory–Transfer Episode: Professor Burgers, Scotophobin, and the (Not–so) Fanatical Fringe. *In:* Cheiron: The International Society for the History of Behavioral and Social Sciences, Berkeley, CA

Stewart, W.W. 1972. Comments on the chemistry of scotophobin. *Nature* 238:202–209

Tate, D.F., Galvan, L., Ungar, G. 1976. Isolation and identification of 2 learning–induced brain peptides. *Pharmac Biochem Behav* 5:441–448

Terenius, L. 1973. Stereospecific interaction between narcotic analgesics and a synaptic plasma membrane fraction of rat cerebral cortex. *Acta Pharmacol Toxicol (Copenhagen)* 32:317–320

Terenius, L., Wahlstrom, A. 1975. Search for an endogenous ligand for the opiate receptor. *Acta Physiol Scand* 94:74–81

Teyler, T.J. 1976. Plasticity in the hippocampus: a model systems approach. *Adv Psychobiol* 3:301–326

Teyler, T.J. 1980. Brain slice preparation: hippocampus. *Brain Res Bull.* 5:391–403

Thiele, T.E., Van Dijk, G., Campfield, L.A., Smith, F.J., Burn, P., Woods, S.C., Bernstein, I.L., Seeley, R.J. 1997. Central infusion of GLP–1, but not leptin, produces conditioned taste aversions in rats. *Am J Physiol* 272:R726–30

Thompson, R., McConnell, J. 1955. Classical conditioning in the planarian, *Dugesia dorotocephalia*. *J Compar Physiol Psychol* 48:65–68

Thompson, R.F. 1991. Are memory traces localized or distributed? *Neuropsychologia* 29:571–82

Thompson, R.F., Thompson, J.K., Kim, J.J., Krupa, D.J., Shinkman, P.G. 1998. The nature of reinforcement in cerebellar learning. *Neurobiol Learn Mem* 70:150–176

Thorndike, E. 1911. Animal Intelligence. MacMillan, New York

Twomey, S.L., Samson, F.E., Jr. 1972. Tubulin antigenicity in brain particulates. *Brain Res* 37:101–108

Ungar, G. 1963. Excitation. Thomas, Springfield

Ungar, G. 1968b. Molecular mechanisms in learning. *Perspect Biol Med* 11:217–232

Ungar, G. 1970. Role of proteins and peptides in learning and memory. *In:* Molecular Mechanisms in Learning and Memory. G. Ungar, editor. pp. 149–175. Plenum, New York

Ungar, G. 1973. The problem of molecular coding of neural information. *Naturwissenschaften* 60:307–312

Ungar, G. 1976. Biochemistry of intelligence. *Res Comm Psychol Psychiatry Behav* 1:597–606

Ungar, G. 1977. *In:* Discovery Processes in Modern Biology. W. Klemm, editor

Ungar, G., Cohen, M. 1966. Induction of morphine tolerance by material extracted from brain of tolerant animals. *Int J Neuropharmacol* 5:183–192

Ungar, G., Desiderio, D., Parr, W. 1972a. Isolation, identification and synthesis of a specific–behaviour–inducing brain peptide. *Nature* 238:198–202

Ungar, G., Desiderio, D., Parr, W. 1972b. Drs. Ungar, Desiderio, and Parr reply as follows. *Nature* 238:209–210

Ungar, G., Galvan, L., Chapouthier, G. 1972c. Evidence for chemical coding of color discrimination in goldfish brain. *Experientia* 28:1026–1027

Ungar, G., Galvan, L., Clark, R.S. 1968. Chemical transfer of learned fear. *Nature* 217:1259–1261

Ungar, G., Irwin, L.N. 1967. Transfer of acquired information by brain extracts. *Nature* 214:453–455

Ungar, G., Irwin, L.N. 1968. Chemical correlates of neural function. *In:* Neurosciences Research. S. Ehrenpreis and O.C. Solnitzky, editors. pp. 73–142. Academic Press, New York

Ungar, G., Oceguera–Navarro, C. 1965. Transfer of habituation by material extracted from brain. *Nature* 207:301–302

Ungar, G., Ungar, A.L. 1976. Endogenous morphine antagonist in morphine–treated rat–brain. *Fed Proc* 35:309

Ungar, G., Ungar, A.L., Malin, D.H., Sarantakis, D. 1977. Brain peptides with opiate antagonist action – their possible role in tolerance and dependence. *Psychoneuroendocrinology* 2:1–10

Uphouse, L., MacInnes, J., Schlesinger, K. 1974. Role of RNA and protein in memory storage: a review. *Behav Genet* 4:29–81

Volkmar, F.R., Greenough, W.T. 1972. Rearing complexity affects branching of dendrites in the visual cortex of the rat. *Science* 176:1145–7

Wade, J.V., Olson, J.P., Samson, F.E., Nelson, S.R., Pazdernik, T.L. 1988. A possible role for taurine in osmoregulation within the brain. *J Neurochem* 51:740–745

Wallace, C.S., Kilman, V.L., Withers, G.S., Greenough, W.T. 1992. Increases in dendritic length in occipital cortex after 4 days of differential housing in weanling rats. *Behav Neural Biol* 58:64–8

Walters, E.T., Carew, T.J., Kandel, E.R. 1979. Classical conditioning in *Aplysia californica*. *Proc Natl Acad Sci USA* 76:6675–6679

Watson, J., Crick, F. 1953. A structure for deoxyribose nucleic acid. *Nature* 171:737–738

Westerman, R. 1963. Somatic inheritance of habituation of responses to light in planarians. *Science* 140:676–677

Wilkins, M., Stokes, A., Wilson, H. 1953. Molecular structure of deoxypentose nucleic acids. *Nature* 171:738–740

Wilson, D. 1986. Scotophobin resurrected as a neuropeptide. *Nature* 320:313–314

Wojcik, M., Niemierko, S. 1978. Effect of synthetic scotophobin on motor–activity in mice. *Acta Neurobiologiae Experimentalis* 38:25–30

Yamamoto, T., Nagai, T., Shimura, T., Yasoshima, Y. 1998. Roles of chemical mediators in the taste system. *Jpn J Pharmacol* 76:325–48

Zemp, J., Wilson, J., Glassman, E. 1967. Brain function and macromolecules, II. Site of increased labeling of RNA in brains of mice during a short-term training experience. *Proc Natl Acad Sci USA* 58:1120–1125

Zemp, J., Wilson, J., Schlesinger, K., Boggan, W., Glassman, E. 1966. Brain function and macromolecules, I. Incorporation of uridine into RNA of mouse brain during a short-term training experience. *Proc Natl Acad Sci USA* 55:1423–1431

Zippel, H., Domagck, G. 1969. Versuche zur chemischen Gedächtnisüber-tragung von farbdressierten Goldfischen auf undressierte Tiere. *Experientia* 25:938–940

Index